RADAR
How it all Began

RADAR
How it all Began

Jim Brown

JANUS PUBLISHING LIMITED
London, England

First published in Great Britain 1996
by Janus Publishing Company,
Edinburgh House, 19 Nassau Street,
London W1N 7RE

Copyright © Jim Brown 1996

**British Library Cataloguing-in-Publication Data.
A catalogue record for this book is available
from the British Library.**

ISBN 1 85756 212 7

Cover design Harold King

Printed & bound in England by
Antony Rowe, Chippenham, Wiltshire

Acknowledgements

*My special thanks to my wife for encouraging me
to write this book, and also
my granddaugher Julie Brown, for doing all the typing
and correcting spelling errors.*

Contents

List of Illustrations

FIGURE 1

RADIO WAVELENGTHS AND FREQUENCIES

VERY LONG	BELOW 30 KILOCYCLES	VERY LOW
LONG	30-300 KILOCYCLES	LOW
MEDIUM	300-3000 KILOCYCLES	MEDIUM
SHORT	3-30 MEGACYCLES	HIGH
VERY SHORT	30-300 MEGACYCLES	VERY HIGH (VHF)
ULTRA SHORT	300-3000 MEGACYCLES	ULTRA HIGH (UHF)
MICRO	3000-30,000 MEGACYCLES	SUPER HIGH (SHF)
MICRO	30,000-300,000 MEGACYCLES	EXTRA HIGH (EHF)

Radio waves travel at 186,000 miles per second or 300,000,000 metres per second, in the same way as light waves.

A radio wave 1-metre long will have a frequency of 300 megacycles; a radio wave 10-metres long will have a frequency of 30 megacycles; a radio wave 10-centimetres long will have a frequency of 3,000 megacycles.

The world 'cycles' has now been replaced by the word 'hertz' (Hz) for alternating currents, so that 50-cycle power supply is now 50Hz. Kilocycles are now kilohertz (kHz). Megacycles are now megahertz (mHz).

Chapter 1

Wireless by 'a Cat's Whisker'

ALTHOUGH THE WORD 'radar' is in general use today, the term only evolved towards the end of World War II in about 1945-46. A number of different types of sets had been developed from 1936 to 1946 and by the end of this period a general term was required to cover all methods of detection using radio frequencies.

'Radar' is a shortened version of 'radio detecting and ranging'. I never used the word radar on any drawing or specification I made from 1936 to 1946. Phrases such as 'radio direction finding' (RDF) and 'radio location' were used on drawings and specifications in the period 1936-37 but were later dropped for security reasons as I will explain later in this book. The radar stations along the south and east coasts of Britain were known as the Chain Home System (often abbreviated to 'CH system') and the Chain Home Low System (abbreviated to 'CHL system'). Other basic types of radar were: the Friend-Foe Indicator (FFI), Gun-Layer (GL) radar and Coastal Defence (CD) radar. These were the early radar sets from which all others developed. A number usually followed the name and this was the means of identifying the set and its exact use.

My first introduction to the 'wireless' (as a radio set was called up to about 1936) took place when my father brought home a crystal set together with a pair of high impedance headphones (see figure 2). This happened early

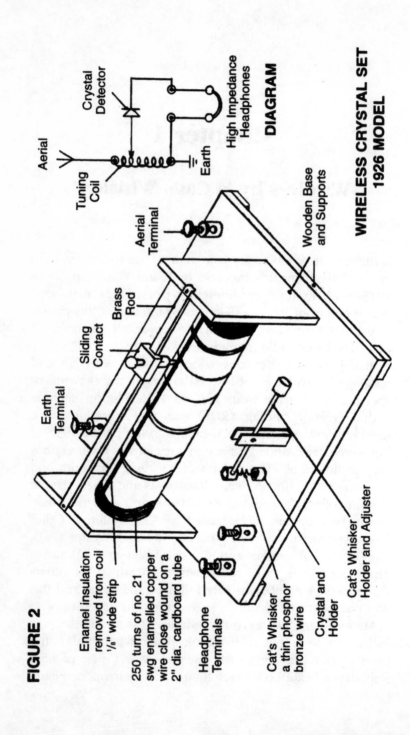

FIGURE 2

Enamel insulation removed from coil 1/8" wide strip

250 turns of no. 21 swg enamelled copper wire close wound on a 2" dia. cardboard tube

Headphone Terminals

Cat's Whisker a thin phosphor bronze wire

Crystal and Holder

Cat's Whisker Holder and Adjuster

Earth Terminal

Sliding Contact

Brass Rod

Aerial Terminal

Wooden Base and Supports

Aerial

Tuning Coil

Crystal Detector

High Impedance Headphones

Earth

DIAGRAM

WIRELESS CRYSTAL SET
1926 MODEL

in 1926 when I was nine years old.

The crystal set consisted of a flat piece of wood on which were mounted two wooden supports for a piece of cardboard tube about 2 in. diameter and 6 in. long. On the tube about 250 turns of thin enamelled copper wire were close wound with a sliding contact to rub along the top of the wire which at this point had a small area scraped off it. This allowed the contact to rub on it and make contact with each turn. The sliding contact was a spring contact out of a bayonet lamp-holder, fitted with a piece of wood for moving the contact along the wire. This was the tuning system. In front of the tuning system were the crystal and the 'cat's whisker'. The crystal was held in a holder – a short piece of brass tube with three small screws set at 120 degrees to each other so that the crystal could be held in position. The brass tube was fixed to the wooden base board and had a wire connected to it. About one inch from the crystal was the 'cat's whisker' with its support and adjusting arm. The support was a U-shaped piece of brass fitted with a round brass arm about $1/4$ in. diameter and 2 in long and secured at the centre by a pin which went through the U-shaped brass support so that the arm could pivot. At one end of the brass arm was fitted a piece of round wood and at the other end was soldered the cat's whisker, which was a piece of very fine phosphor bronze wire formed into a small coil with one end pulled out straight. This was the detector system. In front of the 'cat's whisker' were two terminals marked 'Headphones' and behind the tuner were two terminals – one marked 'Aerial', the other marked 'Earth'.

My father then obtained some wire for the aerial and earth connections and fitted the aerial as high as he could get it in the backyard, drilled a hole in the window frame to bring the wire into the house and ran the earth wire to the water pipe. Now we were ready to test the set. He had written down the instructions to operate the set and he passed them to me:

1.   Connect aerial wire to aerial terminal;
2.   Connect earth wire to earth terminal;
3.   Put headphones on and connect to terminals;
4.   Bring 'cat's whisker' slowly into contact with crystal –
      you will hear a scraping noise in the headphones.
5.   Move slider along the coil and listen for any sound.

Sure enough about two-thirds of the way along the coil it was possible to hear the sound of a man talking. The sound was quite clear. My father had a listen, but my mother would not go near it. I listened to the FA Cup Final played at Wembley Stadium in 1926 on the crystal set.

When a football match was being broadcast, the playing area was divided into eight imaginary squares and when the toss of a coin had been taken to decide which team defended which end, the wireless commentator would say which team was defending the squares numbered one to four and which team was defending the squares numbered five to eight. As the commentator was saying what the players were doing as the game progressed and the ball moved about the field, another voice would come in and say 'square 2', 'square 5', 'square 6' and so on, throughout the match. If the listener had a piece of paper marked out in eight squares, he could follow what was happening on the football field.

When there was a change of programme on the wireless, before the next programme started a voice would say: 'This is 2LO calling.' We kept this crystal set for many years. My mother used to say that I would have hearing trouble as I got older because of wearing the headphones.

I would be 14 years old in February 1930 and I was due to leave school at Easter that year. About October 1929 my father told me that Bill, a friend of his who was a van driver for the Metropolitan Vickers Electric company ('Met-Vick'), Trafford Park, Manchester, had told him that the company was going to hold entrance examinations for their apprentice scheme. To become a trade apprentice I would have to apply for the forms to take the examination. I had never written a letter before so my sister, who was much older than me, told me what to write. I received the forms and again my sister helped me to complete them, making sure that I spelt all the words correctly. I received a reply and was asked to attend in January 1930 on a Saturday morning at 10.00 a.m. I must have passed the exam as I was asked to report to the Education Department at 10.00 a.m. on a date in June 1930. There was another boy there and a man appeared who asked us to follow him. As we walked along he was talking to us and asking questions. He was assessing which one of us should start in the Commissioner's Office. I was left with the Commissioner who explained to me that my job was to take any visitor to see the person they required; I had to telephone the person first to make sure he was free to see the visitor and also to show the visitor the way if he didn't know it.

The Commissioner's Office was on the ground floor at the entrance to the main office block. This was five storeys high and on the top floor was the Management Dining Room. I soon got friendly with Jenny, the lift-operator, who used to tell me when there were any leftovers which required eating. The dining room staff were very good to me and sometimes I took a piece of cake or pie back to the Commissioner.

I was given a clock card and told to clock on at 9.00 a.m., off at 12.00 noon, on at 1.00 p.m. and off at 5.00 p.m., Monday to Friday. On Saturday I was to clock on at 9.00 a.m. and off at 12.00 noon. Everybody at Metropolitan Vickers from the Managing Director to the newly started probation apprentice had to have a clock card or they did not get paid. All staff had to show this card to a man from the salaries department who delivered the wages on Friday afternoons. I was paid 12s. 6d. (62.5p) per week and given one week paid holiday.

Every morning my mother gave me 1s. 2d. (6p) and an

apple and an orange – this was made up of 7*d.* (3p) for tram fares and 7*d.* (3p) for my dinner in the works canteen. I did not tell her that I used to walk or run down Trafford Park and therefore save 1¹/₂*d.* each way. I also received one shilling pocket money per week.

I was still a probation apprentice but part of the apprentice agreement was that I should attend evening classes to obtain some technical qualifications, and so I started going to them in September 1930 at Grecian Street School, Lower Broughton, Salford. The apprentice scheme was organised by the Education Department who kept a check on each apprentice's performance and arranged the movements to the various departments in the factory. We were moved approximately every 9 to 12 months. The time an apprentice stayed in any department was dependent on how he got on with the foreman of that department. Each apprentice spent about two hours per week in the Education Department learning from the instructors about the products of the factory and the ways in which they were used in the various industries of Britain and other parts of the world. Sometimes apprentices also received a pep talk from the management. When an apprentice was moved from one section to another section, he went to the Education Department for an interview with one of the instructors. He would tell the apprentice what the foreman thought about him, point out any evening classes missed and give an assessment of his performance up to date, whether good, bad, or indifferent.

There were three types of apprentice:

1. College Apprentices who came for a year or two to gain practical experience. They were from various universities from all parts of the world and were aged 23 – 27 years.

2. School Apprentices who came from public schools in the British Isles and the Commonwealth for practical experience – they were about 21 – 25 years old

3. Trade Apprentices who were there to learn a trade;

they came from the south-east Lancashire area around
Manchester and were 15 – 21 years old.

I cannot remember any women apprentices in any of the
three categories.

'Met-Vick' Education Department tried to keep in touch
with all ex-apprentices. They issued a book with the names
in alphabetical order with details of apprentices' ages, the
period spent at Met-Vick and how each apprentice had
progressed since finishing his apprenticeship. It also con-
tained the position each one held up to date. The Department sent a letter to the apprentice's last known address
with details of what they were going to put in the next
edition. If it was no longer correct the recipient was
requested to alter it. I still have the last edition published,
dated 1957. I have not received any further request from
the Trafford Park Works of GEC, who have not carried on
the issuing of the records book, probably because it was
too costly to produce.

I had been in the Commissioner's Office for about eight
months when he became ill and died suddenly so that
another Commissioner came. I was due to be moved to
another department but the new Commissioner would not
release me for a further six months because I knew all the
intricacies of the job. Towards the end of 1930 the
depression was having its effect on some of the larger firms
and early in 1931 quite a lot of the probation apprentices
were sacked, including the boy who started with me. The
word 'redundant' had not at this time been used for people,
just for articles. I believe I was not sacked because the new
Commissioner had asked for me to stay with him.

As I got to know the Commissioner he used to tell me
tales about the time when he was in the Royal Navy
(stationed in China) and also about the First World War. I
happened to tell him I was interested in wireless and that I
was thinking of making a three-valve set. He said he had
some spare parts since he had now bought a new wireless
set. He gave me a bag full of wireless parts – there were

valve holders (4-pin), fixed and moving condensers, resistors, coupling transformers, terminals, battery plug connectors, six valves and some wire. The valves had coloured spots on them: red, blue, green, white. The colour indicated whether the valve was an output, a detector, a high frequency (screen grid) amplifier, or a diode. All the valves were suitable for 90 volts DC anode, two volts DC heaters and 9 volts DC grid bias. The only part I lacked was a tuning coil.

From 1922 when the BBC started broadcasting, to the time in 1931 when I decided to build a three-valve wireless set, the development compared with the cat's whisker had been remarkable. By 1931 there were shops selling components, kits of components with diagrams and instructions, complete sets, loudspeakers, and magazines such as *Amateur Wireless, Practical Wireless* and *Wireless World*. All these were offering circuit diagrams, valve characteristics, explaining negative feedback and how it worked and why metal screens and cans reduced interference. I still have a book on wireless which I bought in 1935 for 7s. 6d. (37.5p).

I bought a tuning coil made by Ward and Goldstones of Salford and made an 'L'-shaped wooden chassis for mounting the components. The tuning and reaction condensers' on-off switch and wave-change switch were mounted on the vertical panel; the other components were mounted on the base. The tuning coil and high-frequency valve (green spot) were mounted at the left-hand side (looking at the front of the vertical panel). The detector valve (blue spot) was mounted in the centre and the output valve (red spot) was mounted at the right-hand end. Since we did not have electricity in the house I had to buy a high tension battery (90 volt), a grid bias battery (9 volt) and two 2-volt accumulator-type batteries – while one was in use the other had to go to a shop to be recharged. I still did not have a loudspeaker, but I had the headphones from the crystal set. When the set was switched on, to select a station the tuning condenser had to be slowly turned until a sound was audible. Then the reaction condenser had to be turned from left to right. The volume of sound increased but as it was turned more to the right the sound started to distort and if it was turned still further, the set would go into oscillation and there would be a loud screech through the headphones or loud speaker. Not only did the operator hear the screech but so did every household within about 200 yards of his aerial if they had a wireless set switched on and tuned to the same waveband (i.e. long wave or medium wave). If the set was next door their reception would be blotted out completely. Anybody who built their own wireless sets was accused of doing all the oscillating. I was accused a number of times of oscillating even when my set was switched off! When anyone oscillated and I was listening to my set, I could turn the reaction condenser up and down rapidly about four times. This gave a sound like a present-day police siren. My father bought a loudspeaker and a cabinet for the set which continued to work up to about 1950. When we were connected to an electricity supply in 1936 I arranged a charger for the 2 volt accumulator and bought an eliminator unit which worked from the mains and replaced the 90 volt and 9 volt batteries.

Chapter 2

Apprentice Fitter and Draughtsman

ABOUT MAY 1931 I was told that the company was prepared to accept me as a trade apprentice to become an electrical fitter. My father had to attend and sign the necessary forms. At that time I was still with the Commissioner who was now getting used to the job. I was then moved to the Meter Department. I started at 7.30 a.m. and worked till 12.00 noon, starting again at 1.00 p.m. and finishing at 5.00 p.m. The tram journey from home to Met-Vick took about an hour, so I was up at 6.00 a.m. and got back home about 6.00 p.m. From September to May I attended evening classes at Salford Technical College three nights a week, starting at 7.15 p.m. and finishing at 9.30 p.m., which meant that I returned home at about 10.00 p.m. The reader will see that I did not have a lot of time for other activities.

The Meter Department produced kilowatt hour meters, control panel instruments, ammeters and voltmeters. I had experience on drilling, turning and milling machines. In the 1930s Met-Vick had a bonus scheme based on the Holerith punch card system. A time was given for each job and if you did the job in less time you made a bonus. If you did it in the same time you didn't receive a bonus and if the job took you longer you went into debt. The debt was cancelled the next time you made a bonus. The unions told their members and the apprentices not to make more than 33 $^1/_3$ per cent bonus; if they did the shop steward had a word in their ear. I very soon learnt that a good sharp pencil was the best tool you had. If an apprentice went into debt this was cancelled when he was moved to another department, but it would be recorded at the Education Department. I was never in debt – I had a good pencil and this was part of the training.

I was then moved to the Cooker and Heater Department which had just been started. Met-Vick had developed tubular heaters which they called Red Rings and this section was very busy. They were modern and there was a big demand for them. Up to this time all electric cooker hot plates were solid cast iron and the heating elements were buried inside. They took a long time to heat up and a long time to cool down – they offered no competition to gas rings. These new Red Rings, together with the four heat switches – High, Medium, Low and Off – became very popular because they were much cleaner than gas rings. The heater consisted of two rings, an outer and an inner. On 'High' the switch put the two rings in parallel, on 'Medium' it switched on the inner ring only and on 'Low' it put the two rings in series. The foreman explained this to me just in case I was asked by a visitor. A few weeks after this explanation we had a visit from a section of the Institute of Electrical Engineers and a lady from the party came over to the bench where I was assembling the red rings, switches, fuses and flexible cables on small boiling plates and asked me how it worked. I explained to her exactly as the foreman had told me and she seemed to understand. I got the feeling the foreman had asked her to ask me to check that I had absorbed what he had told me – as far as he was concerned I could not do anything right all through my period in the Cooker and Heater Department. I was there for about one year.

I was then moved to the Transformers Department. As I explained before, after an apprentice was moved and had been in the new department for a few weeks, he was called to the Education Department for a review of his performance in the last department. Since my work did not appear to have suited the foreman of the Cooker and Heater Department I was dreading the day I would be called and was expecting the worst. Sure enough the day came and I went along to the Education Department and into the instructor's office. He was sitting at a desk and open in front of him was a file with a lot of paper in it. He asked me to sit down and asked how I had liked working in the Cooker and Heater Department. I started to say that I had not got on very well with the foreman and was thinking up some excuses for my poor performance. Fortunately he broke in and said, 'That's not what it says here in this report. This is the best report we have ever had for an apprentice.' I was completely floored by this remark and was quick to get out of the office before he changed his mind. I suppose the foreman realised I had some potential and was giving me jobs which were difficult. I was in the Transformer Department for about nine months. They were very busy supplying large transformers for the new national grid system which was being installed throughout the country.

Next I was transferred to West Works where high voltage switch gear was made and again was very busy with work for the national grid. I was put with a man who operated two planing machines, one large, one small. I was to assist him in loading and unloading the large machine and was to learn how to operate the small machine. He was a very good machine-operator and showed me how to grind the different tools to make the various cuts and surface finishes and set the various forward and reverse speeds for cast iron and mild steel frames. At this time the machines were used mainly for switch gear chassis of welded construction. They were about 5 ft high and 2 ft square and we could get five on the large machine and three on the small machine. A one-ton crane was used for loading and unloading. One day I was lifting one of the chassis off the large machine and, as it lifted off the machine bed, part of it caught in one of the clamps and before I could stop the crane it had pulled the chassis out of shape. When the foreman saw it he was not very pleased and said so.

I had been on this job for about three months when one day an instructor from the Education Department came to see me and told me that I was to have a trade test. I expected to be tested on the small planer, but no, he had a piece of metal in his hand and a drawing. He said he wanted me to 'mark it off' in line with the drawing and would be back the following day to see how I got on. I did not have a clue how to start so I showed it to the man who operated the planing machines and told him what was wanted. He said that part of his job was 'marking off' when there was no work for the planing machines, but since I had been there with him there had been a lot of work for the machines and he made more bonus on the machines than on 'marking off'. He then took the piece of metal and the drawing over to a man on the 'marking-off table' and told me to go and see the 'marker-off' as he would fix it for me. The marker-off said he would do it for me and then explain how it was done.

I returned to him as agreed and sure enough there was the piece of steel nicely whitened and marked off exactly as the drawing with centre pops where the holes had to be drilled. The plate was about ¼ in thick and triangular shaped; the two short sides were about 4 in long with the corners rounded off and a ½ in hole in each of the three corners. He showed me how to set it up and mark it off on the marking-off table and also how he had checked the diagonal centres using Pythagoras' theorem, which I had done at evening classes. I then copied his figures on a piece of paper and was ready for the instructor from the Education Department. He came the next day and I showed him the piece of steel all neatly marked off and then showed him how I had checked the dimensions with Pythagoras' theorem. He just looked at it and said that his only comment was that my centre pops were a bit big.

When he had gone I went to the marker-off and thanked him for getting me out of a hole and told him to make his centre pops smaller!

In about January 1935 I was moved to the Motor Test Department. I was put with one of the technicians on the direct current motor test bed for motors and generators up to 100 horse power. My job was to grind the carbon brushes with glass paper placed between each brush-holder and the motor commutator, set the brushes in the neutral position, fit a pulley on the shaft and fit a belt between the motor being tested and a generator which was acting as the load. From this we obtained the power developed by the motor and also the temperature rise of the motor windings. One day I had put a motor on test which had a separately excited field winding on 500 volts DC. I forgot the motor was separately excited and when the motor had stopped, I opened the terminal box and put a spanner on the terminals to remove the connections. There was a big flash and I got quite a shock. I was extremely careful after that. I had forgotten to switch off the 500 volt DC separate supply.

My next move, in August 1935, was to the Research Department Drawing Office. I was 19½ years old and once again I worked 9.00 a.m. – 12.00 noon and 1.00 p.m. – 5.00 p.m. I also worked on Saturday mornings. I was still an apprentice and was receiving about £1 5s. 0d. (£1.25) per week. Since I had moved into the Drawing Office I could no longer wear overalls and a battered raincoat so my mother bought me a new suit, two new shirts, a tie, a pair of shoes and a new raincoat. I thought she was being a bit extravagant since I was only expected to be in the Drawing Office for about a year or even less if my work did not suit the chief draughtsman. I still lived in Salford and went into the North Gate at Met-Vick. Unfortunately the Research Department was near the South Gate and was about a seven-minute fast walk or a three-minute run. I could see trouble here because I was not an early bird when I was going to work.

There were ten people in the drawing Office: Mr Starling (the Chief Draughtsman), Mr Taylor (his assistant), Mr Chadwick, Mr Prescot, Mr Bates, two more draughtsmen and two women tracers (whose names I have forgotten), and me. The draughtsmen were aged from 25 to 55 years. Mr Chadwick was the eldest. In the Drawing Office at this time, staff were always addressed as Mr or Miss – the use of first names was just not the done thing. There were very few Mrs – when they got married most women stopped working in factories (1939 changed all that). I do not think they were expecting me because there was no drawing board for me to work on, so I sat at a desk where the standards books resided and was told to have a look at these for the time being. There were three standards books each about 6 in. thick, 18 in. long and 12 in. wide – they never moved off the table as they were too heavy to lift. Using these books it would have been possible to design a power station. After about two days with the standards books I was fixed up with a drawing board near Mr Chadwick as he had more room than the others. I spent the first two months doing minor modifications on drawings from marked-up prints and getting to know the system of ordering materials and parts, obtaining and recording drawing numbers and specifications. Mr Chadwick was a great help and explained to me how not to upset Mr Taylor.

The first job I had to design on my own was an engine indicator unit. These units were to be used in conjunction with Esso petrol and Rolls-Royce motors. During the early 1930s the petrol companies and motor manufacturers had been trying to increase the speed of petrol-driven engines by advancing the ignition setting. In this they succeeded but the engines developed a loud knock similar to diesel engines. The engine indicator which I designed was arranged with a contact breaker which could be advanced or retarded relative to the piston top dead centre. The cam on the contact-breaker was driven from the crankshaft of the engine. In the engine cylinder head, as well as the spark plug hole another hole was drilled. It was a threaded hole 18 millimetre diameter (similar to the spark plug hole), into this additional hole was screwed a pressure plug. The pressure plug had a thin metal end which went into the ignition chamber. Pressing against the metal end were carbon discs and pressing on these was an insulated plug with a metal connection to the carbon discs. These outputs were connected to a cathode-ray tube and by putting additives in the petrol and altering the point of ignition the amount of knock was indicated on the Cathode Ray Tube. The result of all this was the emergence of Esso ethyl anti-knock petrol in red. They dyed it to distinguish it from other petrol and it was available in the period 1936-37. This was probably the first time that lead was added to petrol; we are now trying to get rid of it after 55 years.

As well as being my first design using a cathode-ray tube, it was also my first clanger on a drawing and as every draughtsman will know, you always remember your first mistake. The engine indicator's main body was a cast brass housing for the contact-breaker and cam operating shaft. On the front was mounted a 360 degree protractor with a pointer from the contact-breaker to show the advance or retard of the ignition point. The contact-breaker position was adjusted by a knob which came out on the right-hand side through a boss on the casting. At evening classes when we were doing mechanical drawings we used English projection, but Met-Vick used American projection (now called 'third angle projection') because it was originally called 'The British Westinghouse'. When I started work in 1930 the original name could still be seen in a number of places. It became Metropolitan Vickers during World War I and had kept a lot of the American systems. I had no excuse for doing the drawing in English projection because the standards book went to great lengths to explain the type of projection to be used on drawings. The pattern-maker would assume the drawing was American projection. The net effect of all this was that we got ten brass castings for the bodies of the engine indicators with the boss on the left-hand side instead of the right-hand side. Mr Taylor had a field day when he found out, but Mr Chadwick said it was Mr Taylor's fault because he had not checked the drawings properly. Fortunately the casting surfaces were very rough so we arranged to have them machined and made smooth; this removed the boss and we arranged to braze a boss on to the right-hand side.

The Research Department had developed X-ray apparatus using continuous evacuated equipment in which an electron beam was accelerated and diverted at 90 degrees by the anode at a potential of 250,000 volts DC. The X-rays generated came out through an aperture and could be directed at diseased tissue. A number of hospitals were supplied with the X-ray machines for which I did the drawings. Other equipment I was involved with included electro-magnetic crack detectors. I designed units for valve springs, gudgeon-pins, connecting rods, crankshafts, valves and 12-foot lengths of steel bar. The electro-magnetic crack detectors contributed to the more reliable performance of aeroplane engines during World War II. I was working on the design of equipment for testing the viscosity of insulation varnishes when Mr Starling came along and asked me what I was doing. I explained how the device was supposed to work and he seemed quite pleased. When I finished the explanation he said, 'When you have completed this job, I want you to work with Mr Chadwick and take instructions only from him – he has got some very interesting equipment to design.'

I had become very friendly with Mr Chadwick – he was an ardent Socialist and at 19 years old I had Socialist tendencies. Most of the others in the office were Liberals or Conservatives. Moreover, he was teaching mathematics at an evening class in Stretford, Manchester, and I helped him with some of the awkward calculations. He never understood simultaneous equations. When I told him I had to work with him he was very pleased and said that my first job was to help him design an electron microscope for the Physics Lab. The customer was Manchester University.

The main body of the microscope was a brass tube about 5 ft high. The top part was 4 in. diameter and 12 in. long and this housed the filament to generate the electrons, the anode plate with a very small hole in the centre, the X and Y deflector plates and the sample holder. Below this the body went up to 8 in. diameter, 24 in. long and held the second target position. At the target positions sensitised plates could be inserted so that a permanent record could be kept of the sample magnified 10,000 or 50,000 times. We also designed a high-speed drum which held sensitised film to give a series of records. The rotating drum was driven by an AC induction motor resembling a modern canned pump. The squirrel cage rotor was arranged in the evacuated chamber with the motor stator on the outside. The whole of the electron microscope was continuously evacuated with an Edwards high vacuum pump.

Mr Chadwick and I had almost finished the design for the electron microscope when he told me he had been asked to undertake some secret work which would entail leaving the Drawing Office. He could take another draughtsman with him and, if I wanted to go with him, he would arrange it with Mr Starling. He said we would be working in the Valve Laboratory for about six months but he could not tell me what we were going to do because he did not know. He could have asked for any other draughtsman because they were all more experienced than I, but he had put his faith in me so I agreed to go with him. He said an office was being arranged for us and that we could be moving in about a week. This took place in February 1936 when I was 20 years old on a salary of £1 10s. 0d. (£1.50) per week. Ever since I had been in the research Drawing Office the Valve Lab had always been a bit of a mystery. The doors were always locked and the people who worked there used keys to go in. There did not seem to be much going on – nothing seemed to go in and nothing came out.

Chapter 3

Moved to the Valve Lab – 'Top Secret'

WHEN MR CHADWICK and I went to the Valve Lab to start work we had to knock on the door and we were greeted by Dr Dodds who was in charge of the Lab. He introduced us to Mr Ludlow and a technical assistant whose name I have forgotten. Dr Dodds was a physicist aged about 31 and as I got to know him I realised how technically sound he was. Mr John Ludlow was an engineer aged about 32. Whilst he was quite good, technically he was not as good as Dr Dodds and judging by remarks he made as I got to know him I think he accepted that this was so. The technical assistant's job was to assemble equipment, put it on test and record the results. Mr Chadwick and I were given keys to the doors and told not to discuss the equipment or what went on in the Lab with anybody outside. We were also told that we would have to sign forms relating to the Official Secrets Act. This all sounded very mysterious.

I soon found out what all the secrecy was about. Over the past two or three years Dr Dodds and Mr Ludlow had developed a series of continuously evacuated, demountable, water-cooled transmitting valves which could generate very high frequencies, and very short wave currents suitable for wireless (radio) transmitters. These enabled very powerful radiations from the aerials to be transmitted over long distances. This was a completely new development and was a great advance on valves and the transmitting techniques in 1936. The valves, ranged from 5 kW to 60 kW, were very compact and did not take up a lot of space, even with the vacuum pumping equipment which was part of the transmission system.

There were quite powerful transmitters operating on long wave (30-300 kilocycles) 10,000-1,000 metres and medium wave (300-3,000 kilocycles) 1,000-100 metres but these transmissions were achieved by using a number of large glass valves mounted on removable chassis to develop the required power to the transmitting aerials. For every valve used to deliver power to the aerial there was a stand-by ready to be switched into the circuit to maintain continuous transmission. Consequently transmitting stations were quite large. The GPO was trying out transmitting stations operating on short wave lengths (3-30 megacycles) 10-100 metres using some of the valves developed in the Valve Lab. Amateurs could also use the short wave band. The most successful valves were the short wave valves which we called the 'SW5 valves' and a larger tetrode valve (screened grid) rated at 60 kW output. The SW5 valve was unique because it had one filament, two control grids, two anodes and was water-cooled. It could be arranged as a push-pull oscillator operating at very high frequencies. These valves had been made mainly from sketches and verbal instructions from Dr Dodds and Mr Ludlow. All the pieces of paper were passed to Mr Chadwick with instructions that proper drawings and specifications for the valves should be made so that we could get the parts made in the research workshops, in the various machine shops in the factory and by outside manufacturers.

Dr Dodds told us that if we were in any doubt at all regarding the sketches we were to ask him or Mr Ludlow. We were to ensure that the drawings and specifications were correct in every detail because some of the valves might be required very soon. He did not realise how right he was. Mr Chadwick and I set to work on the drawings. It was decided at the outset that he would do the assembly

FIGURE 3

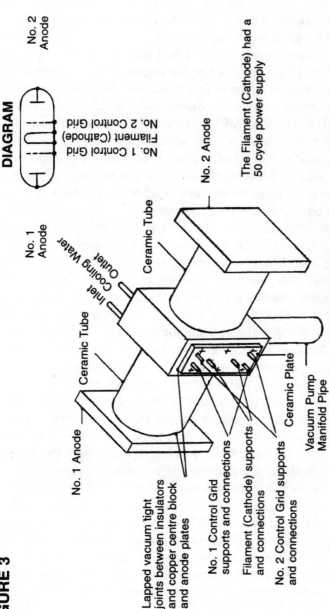

DIAGRAM

No. 1 Control Grid
Filament (Cathode)
No. 2 Control Grid

No. 2 Anode

No. 1 Anode

No. 1 Anode

No. 2 Anode

Inlet
Cooling Water
Outlet

The Filament (Cathode) had a
50 cycle power supply

Ceramic Tube

Ceramic Tube

Ceramic Tube

Lapped vacuum tight
joints between insulators
and copper centre block
and anode plates

No. 1 Anode

No. 1 Control Grid
supports and connections

Filament (Cathode) supports
and connections

No. 2 Control Grid supports
and connections

Ceramic Plate

Vacuum Pump
Manifold Pipe

SW5 VALVE. CONTINUOUS EVACUATED, DE MOUNTABLE, WATER COOLED

drawings and I would do the detail drawings and specifications. When I had looked at the sketches I realised that the valves were just larger versions of the glass valves with coloured spots which I had used in the 3-valve set I made in 1931. The SW5 valve (figure 3) was different. It had a common filament (cathode) and on each side of the filament was arranged a control grid with an anode at each end. The complete SW5 valve was horizontally mounted.

The valve consisted of a solid copper block about 6 in. x 6 in. x 4 in. Machined in one of the 6 in. x 4 in. faces was a slot into which went the filament and two grid assembly. The 6 in. x 6 in. faces were machined out about $4^1/_2$ in. diameter and grooved to take two ceramic tubular insulators which were about 4 in. outside diameter and 4 in long and $^3/_4$ in. thick. On the end of each insulator was the anode which was a copper plate $^1/_2$ in. thick and 6 in. diameter also machined to take the ceramic insulator tube. The filament and the two control grids were mounted on a ceramic insulator 6 in. x 3 in. x $^3/_4$ in. thick with terminals brought through the ceramic for connections. A hole about 1 in. diameter was drilled in the bottom of the copper block into the filament chamber so that the air could be drawn from inside the valve when mounted on the vacuum pumping plant. Where the ceramic insulators joined the copper block and copper anodes these joints were lapped and also sealed with 'appeason grease' to ensure the joints were vacuum tight. What was left of the copper block was drilled with $^1/_4$ in. diameter holes with suitable plugs to form paths for cooling water to circulate. When the valve was not on the vacuum pump it had to be clamped so that it would not fall apart. When it was put on the vacuum pump and the pump started, the clamps could be removed and the valve was held together by the external air pressure. The filament was made from tungsten resistance wire and operated at 20 volts 30 amp AC. The grids were made from nickel-chrome resistance wire. This was the valve that would generate radio waves at 30 megacycles (10 metres) which was to be the basis for the first operational radio direction finder, later termed radar, ever made.

The tetrode valve (figure 4) rated at 60 kilowatts was a much larger valve. It was about 20 in. high and 10 in. diameter. Its base was a copper disc about 10 in. diameter and 1 in. thick and was machined to take the ceramic insulating tube and for the filament, control grid and screened grid assembly which was mounted in the centre of the base plate. There were also holes for the connections to the filament and grids and holes through the base plate into the filament chamber so that air could be drawn from the chamber to evacuate the valve. The ceramic tube was about 8 in. diameter, 1 in. thick and 18 in. long. On top of the ceramic tube was the anode which was a solid copper disc 10 in. diameter 1 in. thick and machined for the ceramic tube and for cooling water tubes. The filament and grid assembly was about 14 in. high and was mounted on a brass centre support rod of ½ in. diameter. The filament was arranged in four vertical loops; around the filament was the control grid and around this was the screened grid. It looked like an elongated beehive. Joints between the copper base and anode and the ceramic tube were lapped and sealed with appeason grease to ensure joints were vacuum tight. The filament was made from tungsten resistance wire and operated at 20 volts 120 amp AC. The grids were made from nickel-chrome resistance wire and the anode voltage was 40,000 volts DC (max.). These were the valves which amplified the output from the SW5 valve to feed the aerials used on the first operational RDF (radar) transmitters for the Chain Home (CH) defence system.

I now began to understand what Dr Dodds, Mr Ludlow and the technical assistant had been doing for three years. Dr Dodds and Mr Ludlow had been making sketches of the various parts for the valves, obtaining the parts, recording the test results and making modifications until the valves were working correctly. The technical assistant had been lapping the joints, assembling the valves and putting them

FIGURE 4

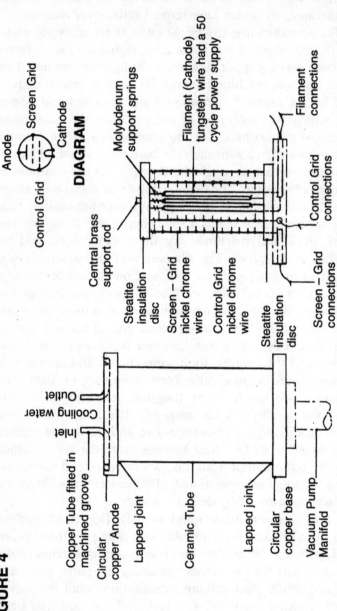

Anode
Control Grid
Screen Grid
Cathode

DIAGRAM

Molybdenum support springs

Filament (Cathode) tungsten wire had a 50 cycle power supply

Steatite insulation disc

Screen – Grid nickel chrome wire

Control Grid nickel chrome wire

Central brass support rod

Steatite insulation disc

Filament connections

Control Grid connections

Screen – Grid connections

Copper Tube fitted in machined groove

Circular copper Anode

Lapped joint

Cooling water
Inlet
Outlet

Ceramic Tube

Lapped joint

Circular copper base

Vacuum Pump Manifold

TETRODE (SCREEN-GRID VALVE) CONTINUOUS EVACUATED, DE MOUNTABLE, WATER COOLED.

on the test bed. This was still going on.

The valve test area was a wire cage with the door interlocked with the isolating switch. There was a vacuum pumping machine and various transformers and rectifiers for the filament, grid and anode supplies and test instruments mounted inside the cage. The valve to be tested was mounted on the vacuum pump, the pump was started and the valve joints were tested for leaks. Connections to the filament, grids and anode were made and the wire cage door and isolating switch were closed. The voltage applied to the filament, grids and anode could be controlled from outside the cage and by varying these voltages and recording each change, the characteristics of each valve could be produced. Dr Dodds had made one valve for demonstration purposes. He had fitted a carbon rod on top of the anode and another carbon rod was fitted to the anode supply voltage lead with resistors to control the anode current. The gap between the carbon rods was about 2 in. and could be increased from outside the cage. When the power was switched on an arc would strike across the two carbon rods and as the filament current was increased the arc was elongated to about 6 in. Dr Dodds had an electrical turntable for playing records at 78 rpm and through a resistor-condenser network, he was able to feed the output of the record played to the control grid of the tetrode. The arc acted as a loudspeaker and the sound produced was excellent and without distortion. When the volume was turned up everything that was loose would rattle – modern ghetto blasters had nothing on this!

The records were all of classical music – Beethoven's Fifth Symphony, Handel's *Water Music*, Grieg's *Peer Gynt*, Strauss waltzes. They never put on Victor Silvester's strict tempo dance music which I was more used to. Mr Ludlow used to refer to this set-up as 'Dodds' All Talking, All Singing Arc'. The technical assistant was a good chap except in one respect – when he was talking every other word was a swear word. Mr Chadwick was a bit religious.

He never swore and did not like to hear other people
swearing so he kept out of the technical assistant's way. If
we wanted to obtain any information from the technical
assistant he asked me to go and see him and if we heard
him coming towards our office Mr Chadwick would lock
the door and say 'Keep quiet and he'll go away.' We had
been making the drawings and specifications for a number
of different valves and during this period Dr Dodds had
been away on and off for a few weeks. Mr Ludlow said he
was trying to get orders for the valves and other connected
equipment. During one of his return visits he asked us how
we were getting on with the drawings and whether we
were in a position to start placing orders for the parts. We
told him we were, but we would require a series of
drawing numbers for reference on each detail drawing,
each assembly and each final assembly drawing. We had
about 20 drawings which required numbers. Dr Dodds said
this was being looked into and he would let us know when
it had been decided which series of drawing numbers to
use.

Chapter 4

Design for the First Transmitters for Chain Home Defence System

ABOUT THE MIDDLE of May 1936 Dr Dodds came to see Mr Chadwick and me and said that he had received an order for four transmitters using some of the valves developed in the Valve Lab and also one for control equipment and spares. We would therefore have to start the design immediately. He went on to say that we were to have an extension to the Valve Lab to assemble and test the transmitters and that a lapping machine had been ordered, but this was to be sited in the Research Workshop. He explained that our work had priority. In the first instance we were allocated 100 drawing numbers. We would have to keep our own records of drawing titles and numbers and we would be given a fireproof cabinet to store the original drawings. This was the beginnings of work on the first operational RDF (radar) transmitter.

When Dr Dodds had gone Mr Chadwick said that whatever it was we were going to design must be very secret because all original drawings in the factory were filed in a central file; only blueprints were kept in the various drawing offices. Originals could be obtained for modification but these had to be requested on a standard form and signed by the chief draughtsman or his assistant. To obtain blueprints off our originals we had to take them personally to the print room, obtain the required number of copies and one for our office file which could be used by Mr

Chadwick, myself, Dr Dodds or Mr Ludlow and which was stamped 'office copy'. It was most important that the original be brought back and filed in our filing cabinet. At first this caused some trouble with the girls in the print room – because all other originals were filed in the central file, they tended to keep the originals. This problem was soon solved.

When we issued a blueprint for parts to be made, we had to cut off the title of the drawing and just refer to the drawing number and item number. This also applied when we placed orders for items to be supplied by outside firms. When the parts be made and delivered, we requested that all blueprints were returned to us in the Valve Lab. These blueprints were then put into an incinerator. In this way we were able to keep track of all information and drawings which we produced and also had a good idea how the work as a whole was progressing. Since I was still an apprentice I was delegated to go for the blueprints and keep records of where they went to and cross them out of the record book when they were incinerated. Now that we had a series of drawing numbers, Mr Chadwick and I could put titles and numbers on the drawings we had produced for the valves. About the first week of June 1936 Dr Dodds and Mr Ludlow came into our office with some sketches. Dr Dodds said he would explain to us with the aid of the sketches what was required regarding the four transmitters but he did not say who they were for. Although all four were basically the same, they were different in some respects, which he would go into later. From the sketches and his explanation it was possible to see that each transmitter was about 8 ft high, 8 ft wide and 8 ft deep and was made mostly of brass angle and sheet brass. Two of the transmitters would be on air and the other two would be on stand-by. There were to be two control desks, one for each pair of transmitters. We also had to design these – they would be made of sheet steel.

Each transmitter had separate compartments for the various functions. There were three valves of the types we had been drawing: one SW5 valve and two tetrodes – the largest rated at 60 kilowatts on continuous wave. In addition there was a vacuum pump chamber, a fixed condenser mechanism and a power source chamber. There were two smaller chambers which could be used at a later date and on the top would be eight aerial feeder outputs. Each transmitter could operate on four pre-set frequencies. Changeover from one frequency to any of the other three frequencies was to be carried out within five minutes. All access doors to the various chambers had to be interlocked with the main isolating and earthing switch. On the front panel of the transmitter would be mounted the wave change control mechanism, variable condensers, tuning adjusters, ammeters and voltmeters. The control desk was to be about 5 ft long, 3 ft deep and as high as a normal desk. It had to be arranged with controls for the valve filaments with ammeters, voltmeters and a 6 in. cathode-ray tube. Terminals had to be provided for connections to the two transmitters being controlled from the control desk.

When Dr Dodds had finished his explanation and had handed over the sketches to Mr Chadwick he asked us if we had any questions regarding what he had explained. I was completely confused trying to take in all this information in such a short time, but Mr Chadwick had no doubt there would be a lot of queries when we got down to details. Dr Dodds then went on to say we could start on the drawings and specifications now and any instructions from the Valve Lab would take priority in the Research Department, the workshops and two other machine shops which were without much work of their own. If there was any hold up in getting work done or in obtaining materials or anything that would affect the progress towards completion of the order, we were to tell him without delay. Mr Ludlow would give us advice and answer any of our questions. Mr Chadwick was more confused than I was regarding the purpose of these transmitters and said he did not know what they were to be used for.

At this stage we did not know what wavelength they were to work on, so it was possible they could be used for broadcasting on some wavelength yet to be decided by the powers that be. Whilst I realised that the transmitters were very powerful I could not understand why there were four wavelengths capable of being changed at very short notice. We decided to get on with the drawings and not ask too many questions. Dr Dodds also told us to order 50 per cent spares, increasing that to 100 per cent spares of parts that could be broken or cracked such as insulators, filaments and grid assemblies for the valves. With the sketches which had been passed to us was a list of metals and insulation materials we could use and those we could not use in the various chambers. In the valve chambers we were allowed to use non-ferrous metals: brass, copper, aluminium, phosphor and bronze. Insulation materials which could be used were pure mica, ceramics, steatite (soapstone), glass Pyrex and calit (which neither Mr Chadwick or I had heard of). We could not use steel, even non-magnetic steel, cast iron or any steel alloy. The insulation materials we could not use were bakelite paper board, bakelite cambric board, moulded bakelite, insulators with an asbestos base and any rubber/cotton laminated insulators. In the other chambers we could use any of the materials listed or any other which we preferred – if in any doubt we had to see Mr Ludlow.

Dr Dodds said he preferred calit as the insulating material in the valve chambers. Calit is a white marble-like material from Germany which can be ground to shape and drilled. Dr Dodds suggested we should make the drawings for the calit insulators and get them on order. Mr Chadwick decided that it would be best if he did the drawings for the cubicles and I looked into the calit insulators, the variable condenser mechanism and wave change mechanism and produced the drawings for them. Most of Mr Chadwick's experience had been on high tension switch gear cubicles similar to the type I had pulled apart with the crane. He was a very good draughtsman and the design and specification of the cubicles and control desks were what he was best at doing.

We had just got going on the drawings for the transmitters when Mr Chadwick was called to see Mr Starling who was still in charge of us. When he came back he said that the Education Department had sent a request to transfer me and that Mr Starling would like to see me. He also said that he would like me to stay in the Valve Lab with him and he was certain that Mr Dodds and Mr Ludlow would also like me to stay. However, if I wanted to move he said that he wouldn't stand in my way. I had been in the Research Drawing Office for about a year and as far as the Education Department was concerned I was due for a move. I went along to see Mr Starling and he said he had a request from the Education Department for me to transfer to another section but it could be ignored – he would like me to stay in the Research Drawing Office. I asked him where I was going to be transferred to and he said a new section was being formed to design projectors (searchlights) and automatic control equipment in another part of the factory and I would most likely finish my apprenticeship there and become part of their staff. He then said he would like me to become one of his staff in the Research D. O. but for the present I would remain in the Valve Lab with Mr Chadwick.

I thought it was very good of him to offer me a job when I was 20 years old, since there were still about three million unemployed. I agreed to stay and went back and told Mr Chadwick who was very pleased. Dr Dodds and Mr Ludlow congratulated me, but when I told the lab technician, as well as congratulating, he added a few swear words for emphasis!

Each of the four cubicles had a front panel 8 ft x 8 ft x $1/8$ in. thick made of mild steel. Behind the front panel were housed the wave change mechanism and the power transformer which was fitted on top with two GES lamp-holders. Into these were screwed two large glass half-wave rectifying valves. These took up about 2 ft 6 in. of the depth of the cubicle and had a steel door at each end which could be bolted for safety reasons. The framework was angle iron. The rest of the cubicle was made of ³/₁₆ in. thick brass plate arranged on a brass angle framework. At about 2 ft 6 in. from the floor was a shelf which went the full length and breadth of the cubicle. There was also a vertical brass panel which went from the floor to the top of the cubicle, set at about 3 ft from the left-hand side. The top and bottom was ³/₁₆ in thick brass plate. This arrangement gave five main compartments: the front one for the wave change mechanism and power pack was 8 ft x 8 ft x 2 ft 6 in.; the lower left compartment for the vacuum pump and manifold for three valves was 5 ft 6 in. long, 2 ft 6 in. wide and 3 ft deep; the upper left compartment for the SW5 valve was 5 ft 6 in. wide, 5 ft 6 in. high and 3 ft deep. There were two smaller compartments in here for equipment on which Dr Dodds was to give us the information later. The lower right compartment was 5 ft 6 in. long, 2 ft 6 in. high and 5 ft deep and housed the fixed power condensers and discharge resistors. The upper right compartment was 5 ft 6 in. wide, 5 ft 6 in. high and 5 ft deep and housed the variable condensers for tuning, wave change switches, two tetrode valves and connections to the eight insulators set on the top of the cubicle, which in turn went to the feeders and then on to the half wave aerials.

There were four doors on the left-hand side of the cubicle which gave access to the vacuum pump chamber (two doors) and two further doors for access to the SW5 valve chamber. There was a similar arrangement on the right-hand side. All the doors were fitted with copper gauze ventilation holes about 10 in. diameter. At the points where all four doors came together a lock had to be arranged. Firstly all four doors had to be locked using a single key and when the doors were locked the key had to be removed and fitted in a lock on the isolating-earthing switch. This applied to both sets of four doors so two keys had to be used to release the isolating-earthing switch before it could be operated to allow the transmitter to be put on the air. The isolating-earthing switch was used for switching the main transformers on and off and also for discharging resistors. The condensers were 1-farad oil-filled working at 40,000 volts DC. The condensers I had used on my three-valve wireless set were a few micro-farads working at 90 volts DC. I ordered eight calit panels for the SW5 chamber – they were about 18 in. x 12 in. x 1 in. thick. It was decided we would drill the holes for the tuning coils and connections to the SW5 valve and outgoing terminals in our own workshop. I then made drawings for the wave change switch supports and driving shafts which were also made of calit. The switch supports were about 3 in. diameter, 5 ft 6 in. long, ground down to $2\frac{1}{2}$ in. diameter in four places for the fixed contacts. There were two of these assemblies for each cubicle. The moving contacts on the switch were mounted on and driven by a calit shaft about $2\frac{1}{2}$ in. diameter, 5 ft 6 in. long. I ordered 16 calit switch supports and 8 calit $2\frac{1}{2}$ in. diameter shafts for driving the moving contacts all from Germany. Little did we know then that these items would help in Germany's downfall.

At this time the extension to the Valve Lab which was being built was about three times the size of the Lab itself. I was now designing the wave change switch mechanism (figure 5) and also the variable tuning condenser mechanism (figure 6). Mr Chadwick was still designing the cubicle and had ordered large amounts of brass angle and pieces of $\frac{3}{16}$ in. thick sheet brass cut to size for the various compartments and doors as well as steel panels and angle iron for the front compartment. The wave change switch calit shaft was extended by a metal shaft to the front panel and the operating handle was held in position on each of the four positions. The condenser tuning mechanism was a bit more difficult. The fixed part of the condenser was mounted on the tetrode valve anode and was made of $\frac{1}{16}$

FIGURE 5

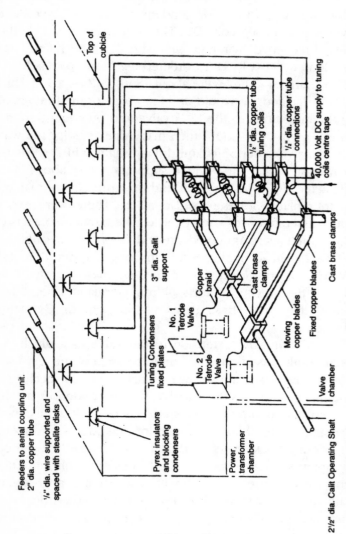

4-WAVEBAND CHANGE-OVER SWITCH MECHANISM FOR TRANSMITTERS CH (CHAIN HOME) DEFENCE SYSTEM

FIGURE 6

4-WAVEBAND TUNING MECHANSIM FOR TRANSMITTERS

CH (CHAIN HOME) DEFENCE SYSTEM

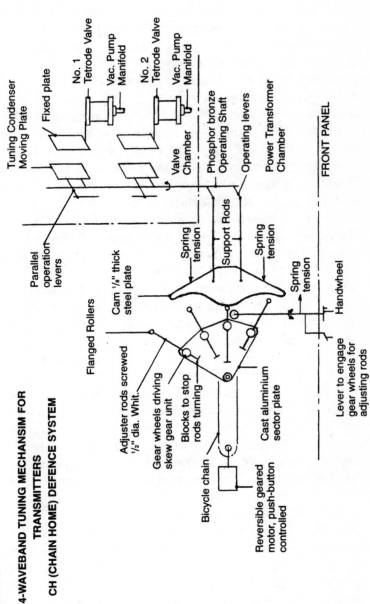

in. thick copper sheet about 20 in. x 20 in. The moving
plate was the same size but had to move in a parallel
motion towards and away from the fixed plate. This was
achieved using a sector plate driven by a geared motor
unit. On the sector plate were mounted four adjustable
arms each fitted with a roller which pushed a spring loaded
cam away or allowed it to return under the spring pressure.
Each adjustable arm could be set in the required position
from the front panel. The cam was connected to the
moving plates of the condensers by brass rods. By
controlling the geared motor the position of the moving
condenser plates could be moved to four pre-set positions
and four pre-set wavelengths were obtainable.

When we started the drawings for the four transmitters
the question arose as to what we were to call them so that
we could put titles on the drawings. Dr Dodds decided we
could call them 'RDF-Tx' which was short for 'radio
direction finder transmitter' and he reminded us to cut off
the title of any blueprint we sent out of the Valve Lab and
also said that the transmitter might be given a Royal
Aeronautical Establishment reference number at a later
date. This was the first hint that the transmitters were to be
used for the location of aircraft. Mr Chadwick and I were
making the drawings for the parts required and ordering
them. Met-Vick had two systems for getting parts made in
the factory and ordering parts from the various stores. This
is where the standards books came into use because they
showed which stores stocked the various materials and
finished parts. The system used for large orders was the
'spec' (specification) system which used foolscap sheets of
paper with Metropolitan Vickers Electric Company at the
top, then came the job title, job number, date, sheet
number and number of sheets. From left to right the
columns were headed: Item Number, Stores/Department,
Number Required, Description and Cost Department. This
system was used when large power stations were being
built but we were not yet in this category. The other system

was called the 'foreman's order'. Foreman's orders were pads of paper arranged with pink, green, blue and white sheets laid out similar to the spec sheets. Each one required a signature; the pads were about half a foolscap sheet in length. The pink sheet went with the drawings to the workshop making the parts, the green one went to the stores supplying the material, the blue one went to the Cost Department and the white one was retained and filed in a folder under the job number. This was the system we had to use to manufacture the first four transmitters. At first only Dr Dodds and Mr Ludlow could sign the foreman's orders but they soon grew tired of that and arranged for Mr Chadwick and me to be able to sign them as well.

It was now coming up to Christmas 1936 and at about 3.00 p.m. on Christmas Eve, Dr Dodds, Mr Ludlow and the technical assistant knocked on our office door which we kept locked, particularly when the technical assistant was around. I opened it and Dr Dodds came in wishing us a Happy Christmas with the other two close behind. He brought with him a box containing sherry, port and whisky and boxes of biscuits, cakes and chocolates. We pulled the drawings to one side and he put the lot on the desks and told us to help ourselves. He was unaware that Mr Chadwick did not drink intoxicating liquor. When the time came to go home at 5.00 p.m. four of us were quite merry and the technical assistant was using words Mr Chadwick had never heard before. Dr Dodds thanked Mr Chadwick and me for what we had done and said we had made a great improvement to the performance of the Valve Lab. I was 20 years old, had just started to learn ballroom dancing and was hoping to buy a second-hand motorcycle in the New Year.

Chapter 5

More Chain Home Transmitters

ON 25TH FEBRUARY 1937 I was 21 years old and had finished my apprenticeship. Overnight my salary went from £1.10s. 0d. (£1.50) per week to £3. 10s. 0d. (£3.50 per week with two weeks' paid holiday. I was still attending evening classes at Salford Technical College – I had a National Certificate in Electrical Engineering and was now taking a Higher National Certificate in Electrical Engineering. At Met-Vick when an apprentice finished his apprenticeship he was called to the Education Department and given a certificate confirming the period he had been an apprentice and outlining what he had done during that period. These certificates were filled in by the tracers in the Research Drawing Office and when they got mine to complete, as well as the official one they made another one which was not very flattering. This was the one that Mr Starling presented to me. I still have both certificates.

The new extension to the Valve Lab was now complete. There was no communicating door between the Lab and the extension for security reasons. A foreman and four fitters had been loaned from the Switch Gear Department and they were sorting out the parts which had already been delivered and trying to understand our assembly drawings. Dr Dodds had given us a sketch showing how he wanted the four cubicles laid out relative to each other and also showing the positions of the control desks. With the help

FIGURE 7

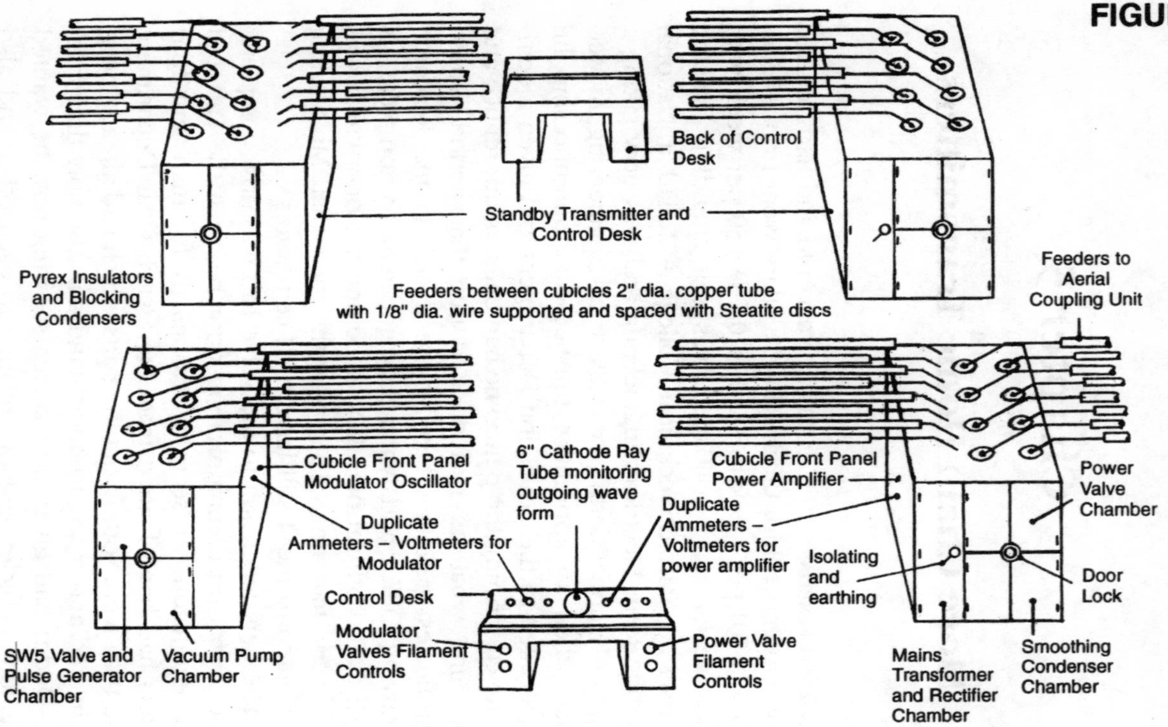

Feeders to Aerial Coupling Unit

Power Valve Chamber

Door Lock

Smoothing Condenser Chamber

Mains Transformer and Rectifier Chamber

Isolating and earthing

Cubicle Front Panel Power Amplifier

Duplicate Ammeters – Voltmeters for power amplifier

Back of Control Desk

Standby Transmitter and Control Desk

Feeders between cubicles 2" dia. copper tube with 1/8" dia. wire supported and spaced with Steatite discs

6" Cathode Ray Tube monitoring outgoing wave form

Cubicle Front Panel Modulator Oscillator

Duplicate Ammeters – Voltmeters for Modulator

Pyrex Insulators and Blocking Condensers

Power Valve Filament Controls

Control Desk

Modulator Valves Filament Controls

SW5 Valve and Pulse Generator Chamber

Vacuum Pump Chamber

LAYOUT OF TRANSMITTERS – CH (CHAIN HOME) DEFENCE SYSTEM

of the foreman I marked out the floor space for each of the four cubicles and the two control desks (see figure 7).

When we first started the design for the four cubicles it was understood that the heavy parts, like the power transformers, vacuum pumping plant, valves and the one-farad condensers would be shipped as loose equipment and assembled on site. Only the cubicle and wave change mechanism would require to be lifted out of our workshop and on to a lorry. Dr Dodds and Mr Ludlow came into our office – if they came together it was usually because something was wrong or because they were having second thoughts about some part of the design. Dr Dodds said he had been asked by the purchaser if the cubicles could be shipped complete to save time in assembling the heavy parts on site. Mr Chadwick said he could not guarantee that the cubicles would not distort and snap some of the calit supports and shafts if they were moved with the heavy parts attached. After some discussion it was decided to build each cubicle on a steel lifting frame. Mr Ludlow and Mr Chadwick would design the lifting frame and Mr Chadwick would produce the drawing and order four frames urgently. This was the first of many modifications and panic design changes which were to follow. I was not involved in this one.

Mr Chadwick mentioned that he was having difficulty understanding Dr Dodds' sketches for the door locks and isolating-earthing switch interlocks. I must admit I too had difficulty interpreting some of the sketches but I usually succeeded with Mr Chadwick's and Mr Ludlow's help. I had assumed that they would be fitting 'Castell' locks which were used extensively for interlocking switch handles and access doors on high tension switch gear and could not understand the difficulty. When Dr Dodds had been in and explained the locking system again to Mr Chadwick and had left our office Mr Chadwick said to me that he still could not understand 'the damn' thing'. I knew he was getting worried that he could not sort it out because he had never used such an expression before. I suggested that I should have a look at the sketches, and told Mr Chadwick that I expected to see a complicated Castell locking system. However, a suitable system using Castell locks could not be found so Dr Dodds had designed one of his own and these sketches were the result. It had been about a week earlier that they had discovered that Castell locks could not be used and Dr Dodds had designed this interlocking system in that time. It was brilliant. When Mr Chadwick passed me the sketches I could not understand them. After about an hour I had some idea how it was supposed to work, so I put a piece of tracing paper on my drawing board and set about copying the sketches. I then cut out the parts which appeared to move and after about three hours I began to understand how it worked.

The locking system was based on standard Yale locks using two locks but only one key. Both locks were ordinary door lock barrels operated by the same key. When using a Yale lock on a door the key is inserted and turned and this turns the barrel of the lock and so releases the door catch. On the Dodds design the key and the lock barrel stayed still and the outer case was turned by a brass knob about 3 in. diameter. Attached to the knob were four catches which engaged the doors. On each door was a peg which pushed a spring-loaded peg on the knob. All four doors had to be closed before the knob could be turned to lock them. The locking process released the key. When the key was taken from the lock barrel the knob could not be turned because the barrel was locked by the tumblers in the Yale lock. A lock was fitted to the four doors on the left-hand side and one on the right-hand side so there were two keys. These keys were now put into similar locks and knobs to those on the doors and when both keys were in position the knobs could be turned to release the main isolating-earthing switch and wave change switch.

I did the drawings for the locking system and ordered twelve pairs of lock barrels and twelve keys (i.e. with 50 percent spare). Some hotels now have this type of lock on bedroom doors. The barrel stays still while the key is inside and when the door is opened the key can be removed. The key cannot be broken in the lock when trying to force the door open.

About March 1937 I bought a second-hand motorcycle for £2. 10s. 0d. (£2.50). Made in 1928, it was a 250cc side valve Matchless, which had acetylene lighting. I soon changed the lighting to electric but could not fit a generator so I had to charge the six volt battery at home. It had a magneto for ignition and had a top speed of about 55 m.p.h. I had to learn all about internal combustion engines very quickly to keep it going. I passed the driving test which had only recently been introduced (I still have my original licence). I was still learning ballroom dancing and now knew the difference between a waltz, a fox-trot and a quickstep. We were still very busy and Dr Dodds kept pressing Mr Chadwick and myself on to greater effort. Mr Chadwick asked me if I was prepared to work some overtime when I finished evening classes in April and I agreed. We started working two nights a week to 7.00 p.m. and sometimes on Saturday afternoon and Sunday. Staff did not get paid for overtime though we were given our tea when we worked in the evening and our dinner on Saturdays and Sundays. I have referred to the break at 12.00 noon to 1.00 p.m. as dinner because at that time, the word lunch was not used by ordinary people. The foreman and the four fitters were also working overtime. The lifting frames had been delivered and the cubicles were taking shape.

In April 1937 the two control desks, which Mr Chadwick and Dr Dodds had designed and ordered from a firm who specialised in this type of sheet steel metal work, were delivered. They were about 5 ft long, 3 ft wide, 2 ft 6 in. high and each had a raised portion at the back. Mr Ludlow was to deal with the control desks and I was to carry out any drawing and ordering which had to be done. Each control desk was divided into a left-hand half and a right-hand half, one half for one transmitter, the other half for the other transmitter. In the centre on the raised back portion was a cathode-ray tube about 6 in. (150 mm) diameter. Also on the raised part were ammeters, voltmeters and various small switches and indicator lamps. In the panels on each side of the knee hole were mounted the control variable resistor units for controlling the filament currents of the SW5 valve and the two tetrode valves in each cubicle. There was also places for mounting small valve amplifiers and the control unit for the cathode-ray tube. At the back of the control desks were rows of terminals for the outgoing cables to the two transmitters. Each terminal was labelled with a number which corresponded with a similar number in the transmitters. The ammeters, voltmeters, switches, indicator lamps and control resistors were also labelled with their functions. The cathode-ray tube and control unit were obtained from an outside supplier and with Mr Ludlow's help I wrote out a specification and ordered four. This gave me 100 per cent spare.

When I received an outline drawing of the cathode-ray tube and control unit and showed it to Mr Ludlow he said he would have to fit a 'mu-metal' tube as far around the cathode-ray tube as was possible to screen it from any stray electro-magnetic field set up when the transmitters and receivers were in operation. I had never heard of mu-metal. Mr Ludlow had to explain that it had very good magnetic properties, much better than mild steel and better than the steel used for electric motors and transformer stampings. It had very high permeability and, since the cathode-ray tube would be fitted with external magnetic deflection coils as well as the internal electrostatic deflection plates, we had to fit the mu-metal tube over the external coils.

It seemed that the cathode-ray tube I was dealing with showed the outgoing waves from the aerials were sending out and this information was fed to the receiver control room which had a 9 in. cathode-ray tube. It was shown as a vertical line at the left-hand side of the cathode-ray tube. Any aeroplane in the path of the transmitted radio waves would reflect the waves which could then be detected by the receiver aerials, amplified and fed as a vertical line to the right-hand side of the 9 in. cathode-ray tube. The time base moves the cathode ray from left to right of the tube so that the distance between the transmitted vertical line and the received vertical line is equal to the time taken for the radio waves to travel from the transmitting aerials to the aeroplane and return to the receiver aerials. All this takes place in micro-seconds. An aeroplane flying 50 miles away would show on the cathode-ray tube as a displacement of the two vertical lines of about 0.00052 seconds; 100 miles could equal 0.00104 seconds. I had now completed all the details for the control desks and ordered all the parts including the mu-metal tubes which were about 4 in. diameter, 8 in. long and $^1/_{32}$ in. thick. In the compartment directly behind the front panel of the transmitter cubicle was a power transformer. This was oil filled and was about 5 ft long, 2 ft wide and 2 ft 6 in. high. It was on wheels and was run in on two channel irons fixed to the floor of the compartment. It was about 150 kVA and fitted on the top were two Goliath Edison screw lampholders into which were screwed two rectifying valves made of glass. The valves were about 24 in. high and 12 in. wide; on top in the middle were the terminals. The transformer secondary winding was arranged to feed the two rectifying valves as half wave rectifiers and produced 40,000 volts DC at 3 amp. The transformer also provided the filament and grid supplies for the SW5 valve, the two tetrode valves and the control desk.

At the points where the 40,00) volt DC supplies went through the metal panels condenser brushings were fitted. The four 1-farad fixed condensers were for smoothing out any ripple on the DC voltages going to the valve anodes and screened grid supplies. My next job was to design and detail the main isolating-earthing switch. This was a shaft about ³/₄ in. diameter and 8 ft long which went from the front panel through the front compartment, through the 1-farad condenser compartment to the back panel. It controlled the power to the main transformer and earthed the 1-farad condensers via fixed discharge resistors when the power was turned off. Since the condensers were at 40,000 volts DC when they were working, the clearance to earth for the discharge resistors and any connections to the condenser live terminals had to be about 3 in.

Dr Dodds came into our office with some more sketches and said that these were for the tuning coils to be fitted in the SW5 valve chamber and also across the four wave change switches. He also said that the correct dimensions for the tuning coils could not be determined until the transmitters had been tested. The coils shown should give a wavelength of about 10 metres but he was not sure. A drawing was required for each tuning coil with a list to record the estimated wavelength and the actual wavelength developed during the test. It was decided that I would make the drawings. The tuning coils for the SW5 chamber were made from ¹/₄ in. outside diameter copper tube each with three turns wound on a 1 in. diameter mandrel. They had a flat copper plate brazed at each end of the coil for termination. I ordered twenty of these so that if any modifications were required they had plenty of spares.

Tuning coils which were to be connected across the wave change switches were made from ¹/₂ in. outside diameter copper tube with three, four, five or six turns and wound on a 3 in. diameter mandrel. Each coil had a centre tap terminal and a flat copper plate brazed at each end. I ordered 6 of each type giving a total of 24 tuning coils, again allowing plenty of spares if modifications were required. I now realised that if these transmitters were to work on a 10-metre wavelength this would mean the oscillators working at 30 megacycles with an output of about 300 kilowatts. This would mean that they were the most powerful short-wave transmitters ever made.

I decided to tell Mr Chadwick what I had worked out from the sketches of the tuning coils but I got no reaction from him. I think all this talk of the valves oscillating at 30 million cycles per second, tetrodes, control grids, screen grids, anodes and cathode-ray tubes was a bit beyond him at his age. The situation would be similar for me if I now tried to understand the technicalities of modern computers – it's a different language.

There were two more jobs which remained to be done: one was the arrangement on top of the cubicles for the outgoing feeders to the aerial coupling unit and the other was the completion of some test equipment which Dr Dodds was designing. Mr Chadwick decided that he would make the drawings for the top of the cubicles and the outgoing feeders and I would do the test gear. The cubicles were progressing and the parts were being assembled and installed as they became available – there were very few modifications to our drawings. Dr Dodds had said that he would like to start testing in May 1937 and ship two cubicles and one control desk in June 1937. We could not test them properly because we had no aerials or receivers so any testing had to be simulated. Here Dr Dodds came up with the solution with some more of his sketches.

This was a unit which held four photo-electric cells arranged in parallel. The cells were flat and about 2 in. square and contact had to be made on the back and along the front edges. The output was a DC voltage proportional to the amount of light falling on the photo-electric cells. A one milliamp full-scale deflection ammeter was connected to the photocell unit; the ammeter at this stage had a blank scale. Six 150-watt tubular tungsten filament lamps were arranged in parallel on insulated brackets and fitted with connections across two of the eight output terminals on top of the cubicle. They were set about 10 in. from the cubicle top and the photocell unit was placed in the centre of the tubular lamp array on top of the cubicle and the ammeter was put on the ground with the two-core cable to the ammeter trailing down the cubicle. We made 20 of the tubular lamp assemblies although only eight were required to test two transmitters. This was to cover for breakage and burn-out of the lamps while tests were being carried out. We also ordered 20 milliammeters for the same reason.

When the transmitters were fitted with all their tuning coils and when the wave-change switch was set to one of the four positions and the set was switched on, the idea was to make the tubular lamps start to glow by moving the power operated variable condensers. As the setting came to the designed frequency, the lamps would glow very brightly and the DC ammeter pointer would deflect in line with the brightness of the lamps. The ammeter could now be calibrated in kilowatts since we knew the current flowing through the tubular lamps and the open circuit DC voltage available.

The test was carried out on all four wavelengths by moving the wave change switch to each of its four positions and setting the variable condensers to give maximum brightness of the tubular lamps. The wave change mechanism for moving the variable condensers was now fixed for that position and could be moved to the next position. This was repeated four times for each transmitter. With this arrangement the output frequency of any transmitter could be changed and within five minutes it could be transmitting on a different frequency. It was very important that a frequency could be changed in a very short time and, if necessary, changed back equally quickly. Thus, when the transmitters were on the air and the receiver cathode-ray tube was showing interference, the transmitter operator could be asked to change the frequency being transmitted with a view to removing the interference. This was one of the reasons for all the secrecy. Mr Chadwick had been making the drawings for the top of the cubicle which had eight Pyrex insulators, feeder tubes from the insulators to the coupling unit and a heat exchanger for the cooling water to the valves. The top cover of the cubicle was made of brass $^3/_{16}$ in. think and in it were eight 7 in. holes. Over each hole was a Pyrex insulator measuring about 8 in. diameter and 6 in. deep with an internal flange and skirt which went through the 7 in. hole. They were a bit like pudding basins but had a hole in the top for $^1/_2$ in. brass bolts which were for the connections to the feeder tubes.

We had arranged the four cubicles in the workshop as they would be arranged on site. The front panels faced each other and were spaced about 15 ft apart. The control desk for each pair of cubicles was arranged so that the operator who was sitting there could see both the front panels of the cubicles. The aerial coupling units would be hung from the ceiling. In the coupling units were air-cored transformer coils, some connected to the aerials and some connected to the output terminals of the cubicles. The feeders from the cubicles to the coupling units were copper tubes measuring 2 in. outside diameter, and down the centre of each tube was a $^1/_8$ in. diameter copper wire supported by steatite discs spaced about 9 in. apart. The steatite discs were held in place by soldering a brass washer on each side of the disc. The $^1/_8$ in. diameter copper wire with all its steatite discs was pulled through the 2 in. copper tube. There were four tubes about 8 ft long from each cubicle to the coupling unit.

The feeder tubes were arranged like a modern co-axial cable but they were much larger. In the period when we were designing these high frequency transmitters it was generally believed that high frequency currents flowed on the surface of conductors. Consequently all tuning coils and connections carrying high frequency currents were made from copper tubing, preferably measuring $^1/_2$ in. outside diameter and certainly not less than $^1/_4$ in. They were so rigid that they did not require any additional supports other than the item they were connected to, and since the connections were not insulated they had to be spaced with at least 3 in. clearance between each connection and any parts at earth potential.

Mr Chadwick had completed the necessary drawings and

specifications and had ordered all the parts required. The cooling water connections between the valves and the heat exchanger were to be rubber tubes wound in a coil like a high frequency choke. Distilled water had to be used and the rubber tubes were fixed with non-ferrous hose clips.

Mr Chadwick had mentioned that Dr Dodds and Mr Ludlow were having some difficulty with the Pyrex insulators and that he had not been able to complete the drawings showing how they were to be fixed in position on top of the cubicle. I could not see any difficulty: there was a $1/2$ in. hole in the top of the insulator and since the connections were rigid the connection would hold them in place. Mr Chadwick then said that there could be no direct connection from inside the cubicle to the terminal on top of the Pyrex insulator.

At first I could not understand the reason for this, but after a short while I realised that this was a blocking condenser to block the 40,000 volts DC from going up the aerials. Mr Chadwick had said that if we could come up with the answer to the problem both Dr Dodds and Mr Ludlow would be very happy since they had no solution. He thought they had not realised there was a problem in the first case. Fortunately the Pyrex insulators had been delivered so I suggested to Mr Chadwick that we get one and bring it into the office so we could have a good look at it and he agreed. It was about 8 in. diameter and 6 in. deep with a skirt which went through the 7 in. hole in the top of the cubicle. It also had an internal flange measuring about 7 in. diameter. I made a disc from some thick card-board about 7 in. diameter and fixed a pencil in the centre with some sticky tape. After about an hour of trial and error I realised that if I cut two small flats on the disc I could manoeuvre the cardboard disc up into the air space of the insulator. Then, by turning the disc, it would land on the internal flange of the insulator and so could hold it in place. I showed it to Mr Chadwick who went along to Mr Ludlow's office and told him we had sorted out his Pyrex insulator problem. When Mr Ludlow saw what I had done and I demonstrated how it worked I thought he was going to kiss me! We had one made from $3/16$ in. thick brass with $1/2$ in. copper tube brazed in the centre of the disc for connection to the wave change switch. The brass disc formed one plate of a blocking condenser; the other plate measured about 8 in. diameter brass on the outside of the Pyrex insulator and was fixed through the $1/2$ in. diameter hole in the top of the insulator.

In April 1937 Dr Dodds had been away for about ten days and on his return he came into our office. After congratulating me on sorting out the Pyrex insulator problem he went on to say that he had received a further order for another sixteen transmitters, eight control desks and spares. We also had to remove all references to 'RDF. transmitters' or 'radio location' on all drawings and specifications and replace them with the words 'CH station'. This was for security reasons and it was to be done as a matter of urgency. We were to scrap all blueprints and ensure they were burned in the incinerator. 'CH station' stood for 'Chain Home station' and we were to be given a Royal Aeronautical Establishment (RAE) reference number for the transmitters. The number was to be given to us later. The total of twenty transmitters, ten control desks and spares was the largest order the Research Department had ever had. We were given a new job number for the sixteen transmitters and were told we had to issue our manufacturing instructions on the specification sheet system and not on foreman's orders. Dr Dodds said that he thought we might require some more help and he suggested establishing another drawing office away from the Valve Lab, since there was no more room. He was quite correct. We had about 200 drawings, mainly 'A' size (that is, 40 in. x 30 in.), files with ordering specifications, bulldog clips hung on the wall holding our copies of the foreman's order labelled: 'In Work', 'Outstanding', 'Waiting Material', 'Outside Supplies', and 'Complete'. In addition there were various catalogues and standards books.

About the middle of April 1937 Dr Dodds came into our office and said he was not very happy with the wave change sector plate assembly. I took out our office copy blueprint and checked all the dimensions but could find nothing wrong so I went along to the new workshop to see the foreman. There were about thirty $1/4$ in. diameter holes in the cast aluminium sector plate which had been marked off and drilled from centre pops. As a result of small inaccuracies in the marking off and the drill running off a bit, when the four adjustable arms were extended to their furthest position they were about $1/2$ in. out relative to each other.

Dr Dodds said that this was too far out but he could accept $1/8$ in. discrepancy. The foreman said he could set the four adjustable arms in their correct position and dowel pin them and would probably have to do the same with the other three sector plates for the other cubicles. If we wanted them to be accurate without dowel pinning, a drilling jig would be the best solution. Fortunately only one sector plate had been drilled and we still had more plates than we required. Dr Dodds said that since we now had twenty cubicles to make instead of the original four we could justify the cost of a drill jig and asked me to go and see the tool room foreman. I had to tell him that the jig was very urgent and he could work as much overtime on it as was necessary.

I took with me a blueprint of the sector plate showing the holes required and a blueprint of the complete assembly to explain how the mechanism worked and to impress on him why the position of the holes needed to be so accurate. I also took one of the cast aluminium sector plates. When I showed him my drawing of the sector plate with all the holes dimensioned from centre lines and spaced at various angles in an arc he was not very impressed. And when I told him we had marked one off and it was not accurate enough he was even less impressed and said so. He then said he would have to put the jig on the jig boring machine because of the accuracy we required. He added that he could not work from my drawings as all dimensions had to be given on X and Y co-ordinates and from one hole, preferably a hole at the centre of all the other holes to reduce any error.

All dimensions had to be to three decimal points of an inch, that is, the machine was accurate to 0.001 in. or one thousandth of an inch. The tool room foreman said that once he had a proper drawing and instructions, if we agreed for his men to work overtime, he would produce the jig in three working days. I made a full size drawing as accurately as I could from the previous drawing and, by use of trigonometric tables, I was able to locate every hole as the tool room foreman had requested on X and Y co-ordinates. The jig was completed in the time the foreman had stated. We had no more trouble with the wave change mechanism and the jig was a good investment for later use.

Two of the cubicles and one control desk were nearing completion. Dr. Dodds, Mr Ludlow and the technical assistant had been moving some test gear from the Valve Lab to the new extension with a view to testing the two cubicles and control desk in May 1937 and shipping them by the end of the month or early June. The other two cubicles and control desk could follow in July. To test the cubicles and control desk they had to be roped off and all the cable connections trailed across the floor. By about the middle of May everything was in position for the test to be carried out. Everything seemed to be going well until the technical assistant turned the isolating-earthing switch to its off position.

There was a loud bang and flash and smoke started coming out of the fixed condenser chamber. When the smoke had cleared and the doors to the condenser chamber were opened, it became clear that the discharge resistors had disintegrated and one of the insulators on the condensers had cracked. Dr Dodds said the condenser might not be completely discharged and ensured they were discharged by touching all the live parts inside the condenser chamber with an insulated earthed probe. Fortunately there was no further damage. We had spare condensers but we required some better discharge resistors. Apparently Dr Dodds and Mr Ludlow had designed the condensers to discharge in about five seconds but this was not long enough. It was decided to change the resistors so as to discharge the condensers in about 15 seconds. By referring to the standards books we obtained some suitable resistors from one of the factory stores.

The new resistors were larger than the previous ones but we were able to fit them in and I modified the drawings in line with this alteration. The cubicles were again ready for testing and there were no further panics, although the new resistors still got very hot for about ten seconds when the earthing switch was operated.

Chapter 6

Installation of the
First Chain Home Transmitters

EARLY IN JUNE 1937 we shipped the first two transmitters and one control desk to Bawdsey on the Suffolk coast. Dr Dodds told Mr Chadwick and me that we must keep records of the date all transmitters left the Valve Lab, where they went to, the nearest railway station, the four frequencies each cubicle was tuned to and copies of test results which he would give to us. The reason for knowing the nearest railway station was to enable spares of modification parts to be sent there to be called for by the personnel on the transmitter site.

The transmitters were usually sited on raised ground up to five miles from the sea. There were four steel masts, each one 350 feet high, facing out to sea. The buildings which housed the CH transmitters were partly underground and later camouflaged. A few hundred yards away were four receiver aerial towers, each about 250 feet high. These were made of wood and near to them were the receiver buildings, also partly underground. The receivers, aerial masts and buildings were being supplied by other manufacturers. We were responsible for the transmitters and the control desks for them.

In early July 1937 we shipped the other transmitters and one control desk to Bawdsey. This was the first Chain Home station to be built. The transmitters were originally designed as modulated continuous wave transmitters,

FIGURE 8

DIAGRAM – TRANSMITTERS – CH (CHAIN HOME) DEFENCE SYSTEM

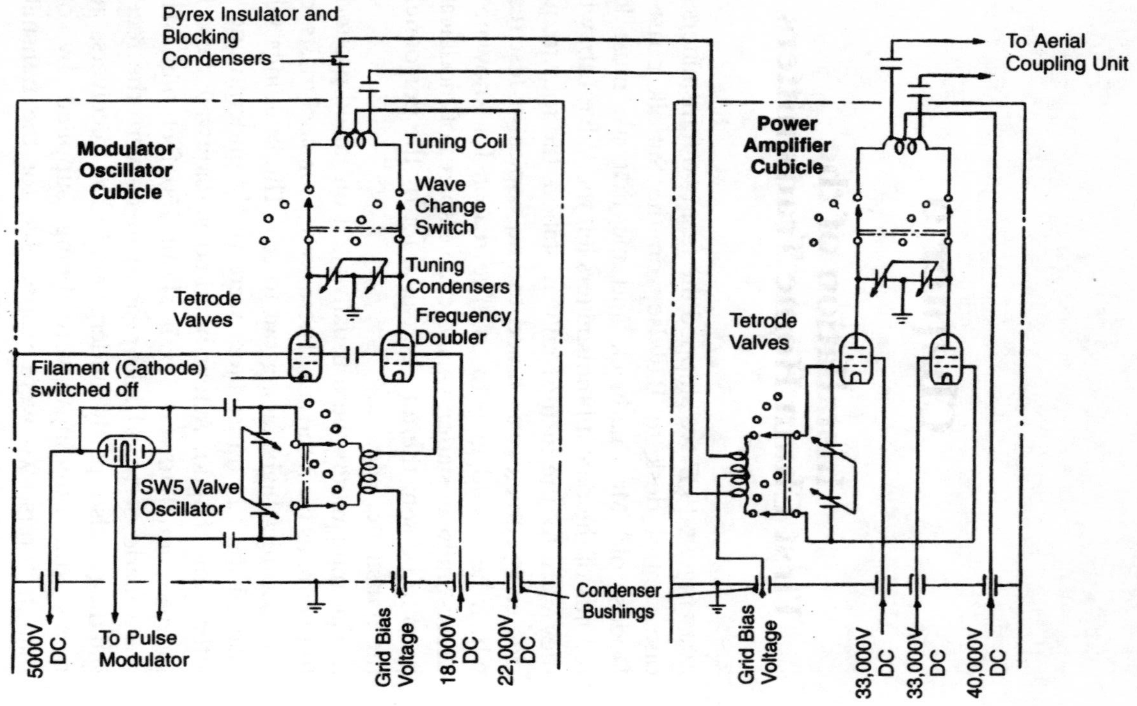

To Aerial Coupling Unit

Power Amplifier Cubicle

Tetrode Valves

DC 40,000V
DC 33,000V
DC 33,000V
Grid Bias Voltage
Condenser Bushings

Pyrex Insulator and Blocking Condensers

Tuning Coil
Wave Change Switch
Tuning Condensers
Frequency Doubler

Modulator Oscillator Cubicle

Tetrode Valves

Filament (Cathode) switched off

SW5 Valve Oscillator

To Pulse Modulator

DC 22,000V
DC 18,000V
Grid Bias Voltage

DC 5000V

similar to a normal broadcast transmitter on 10 to 15 metre wavelength (that is, 30 – 20 megacycles) with an output of 250 – 300 kW. They were the most powerful short-wave transmitters ever built up to that time. (See figure 8.)

Modulation of the carrier wave was achieved by using the 50 cycle per sec. frequency of the national grid. Anyone who had a receiver tuned to 10 – 15 metre wavelength would hear a very loud 50 cycle hum and come to the conclusion that there was something wrong with the national grid system or that there were a few bad connections or broken insulators. The hope was that anyone on the other side of the North Sea who picked up the transmissions in the 10 – 15 metre wavelength would also come to the same conclusion.

We had designed and produced four radio location (radar) transmitters and two control desks in about a year. The others that followed were to have a very great impact on the outcome of World War II which was now approaching. Everything achieved had been mainly by the efforts of Dr Dodds, Mr Ludlow, Mr Chadwick and myself. The first four transmitters were sent to Bawdsey because Mr Watson-Watt had set up an experimental station for the purpose for locating aeroplanes there in 1935. This was as a result of some experiments carried out by Mr Watson-Watt in conjunction with the GPO Radio Transmission Group at Daventry, who had been trying out some of the continuous evacuated valves developed in the Valve Lab at Met-Vick by Dr Dodds and Mr Ludlow.

The trials with the Met-Vick valves started in 1932 and since they were much more powerful and took up less space than the best glass valves produced up to that time, they were being used and further developed for short-wave long distance radio transmissions. Dr Dodds and Mr Ludlow had been involved in the project with the GPO Transmissions Group at Daventry from 1932 to 1935. At this stage the reason for all the secrecy was commercial since a lot of effort and money had gone into the development of the valves. About 1933 one of the GPO engineers noticed that when an aeroplane went past the transmitting centre it affected the radio reception. He must have been a very observant man because there were not many aeroplanes flying about in 1933. He was the real inventor of RDF (radar).

This phenomenon was brought to the notice of Mr Watson-Watt who was head of a radio research station at Slough and early in 1935 he arranged for a trial to be carried out using a Daventry transmitter on 50 metre wavelength (6 megacycles). The trial was partly successful so it was decided to set up an experimental station at Bawdsey using glass valves but still working on about 50 metre wavelength. They were able to locate aeroplanes up to 20 miles from the transmitting and receiving aerials. In 1935 there was a general belief that there was no suitable defence against bomber aircraft even in good daylight and much less in cloudy weather and during the hours of darkness, and that bombers would be able to get through any of the known defence systems. The only defence was to build bigger and better bombing aircraft and retaliate. Mr Watson-Watt and his team very soon found out that a more powerful transmitter working on a shorter wavelength with higher aerial masts than they were currently using would be required. Moreover, both the transmitting and receiving aerials would work better if they were horizontally polarised.

The theory was that a bomber aeroplane had a wing span of about 30 – 40 ft and as the electro-magnetic radio waves struck the wings they would generate eddy currents in the metal structure and the wings would act as a di-pole aerial and re-transmit electro-magnetic waves at the same frequency and waveform as the transmitter was transmitting. Therefore, with a wavelength of 10 – 15 metres and a lot of power being radiated from the transmitting aerials, a very strong signal would be received by the receiver aerials which could be amplified and displayed on a cathode-ray tube. Mr Watson-Watt had been given an allocation of money to develop the radio location system and in May 1936 had given Met-Vick an order for four transmitters and two control desks with specifications based on what had been found out during the experiments at Bawdsey.

We had now shipped the four transmitters and two control desks and had a further four lifting frames in place for the Chain Home station site which was to be decided later. Since we had ordered quite a lot of spares for the first four transmitters we were well placed to start on the next four. We only had space to work on four at a time but had some space to build sub-assemblies which would be fitted in the cubicles later.

We now had to use the specification system to place orders with the various workshops in the factory. The foreman's orders which we had used for the manufacture of the first four cubicles and two control desks were only intended for small jobs. A foreman's order was issued for each operation. For example, one went to the Pattern Shop for a pattern to be made, one went to the Foundry for castings to be made from the pattern, one was sent to a machine shop for any milling or drilling to be carried out. It was not necessary to issue all the foreman's orders at once – they could be issued as and when required.

Using the specification system, the items being made had to be itemised from the raw material to the assembly shop in the correct order, with material specifications drawings and item numbers and paint and electro-plating specifications shown. Items which were bought from outside suppliers had to be included with the indent number and order number issued by the Supply Department on the specification sheets. There were usually about six copies required but there could be up to eight – every department mentioned on each sheet had to have a copy as well as the Cost Section. An office copy was also kept. When an alteration or modification was required to any item, reference was made to the item, the modification required had to be given and instructions as to what to do with any parts already made. For example, it was necessary to indicate if an item should go for scrap or if it should be returned to the stores it came from. In other parts of the factory all this was carried out by typists with carbon copies. The seventh and eighth copies were usually very dim. The seventh copy went to Cost Section and the eighth was used as an office copy which someone went over with a pencil to make it clear. The best copies were sent to the departments doing the work so that there was no excuse for not being able to read the instructions. There were no office copying machines in 1937.

Dr Dodds had suggested that we order about 20 per cent more of the parts which could get broken or damaged and 10 per cent more of the other parts for the sixteen transmitters and eight control desks for which we now had a firm order. Mr Chadwick and I were bringing the drawings up to date and ordering all the parts for the new order. Since we did not have a typist we had to print the instructions on the specification sheets using carbon paper for copies.

Whilst we now had priority in a number of workshops in the factory it was best to get in touch with the people concerned and make sure they knew just how much work they were taking on. This could be done by telephone but I preferred to go along to see the person concerned with a copy of the drawing and to check on parts which were behind on delivery. If there was any delay I just pointed out I would have to tell Dr Dodds. Since all the departments we dealt with had written instructions that they could work as much overtime as was necessary, deliveries were usually earlier than we had requested.

Mr Chadwick did not seem to mind me dealing direct with the departments supplying the parts and writing out the specification sheets. I quite liked dealing with people direct and sorting out problems but Mr Chadwick had seen it all before.

We shipped four transmitters and two control desks to Great Bromley in Essex in October 1937. The people at Bawdsey had carried out some trials with the transmitters we had already delivered and were very happy with the results obtained. Dr Dodds said that we would be getting orders for more transmitters to be made and that there were to be further trials early in 1938 depending on weather conditions. It was now Christmas 1937 and Dr Dodds again arranged a small party for myself, Mr Chadwick, Mr Ludlow and the lab technician in our office. During the party Dr Dodds said that the lab technician had something to tell us. I thought that it was some new swear word we had not heard before, but no, he was leaving the company to take up a position with another electrical company. It would be someone else's turn for an ear-bending from his swear words. Mr Chadwick wished him all the best in his new job, but seemed more interested as to when he was leaving than when he was starting his new job. About a year after he had left us he came to see how we were getting on and on this occasion he didn't use one swear word all the time he was with us. Mr Chadwick said somebody had shown him the error of his ways.

We shipped a further four transmitters and two control desks in January 1938. These went to Dunkirk in Kent (see figure 9 for location of all the CH stations). We also received another order for 68 transmitters and 34 control desks, making a total of 88 transmitters, 44 control desks and spares. In financial terms I believe this was the largest order that Met-Vick had ever had. This made a total of 22 Chain Home stations. At this stage we did not know where they were to be sited – the whole programme was still very secret. We were still keeping our own original drawings and specifications, keeping records of all blueprints issued and ensuring that all used copies were incinerated.

A new office was being established in what had been a physics laboratory on the first floor in the main Research Department Building. The room was much larger than we

FIGURE 9

Map showing the
22 Transmitters sites,
July 1940
CH (Chain Home)
Defence System

really required for the amount of work we now had on hand but it came in handy later as the demand for all kinds of radio equipment increased as World War II progressed. Mr Chadwick had decided that we required more fire-proof drawing cabinets, another draughtsman, access to a typist for the specification sheets, use of a tracer, use of a progress chase and a technical clerk. He told Dr Dodds our requirements. I think Dr Dodds would have agreed to anything we requested so long as the transmitters were progressing and he was meeting delivery dates.

An area in the West Works had been partitioned off for assembling the cubicles as the programme consisted of twenty cubicles and ten control desks. Mr Chadwick asked me if I thought any of the draughtsmen in the Research Drawing Office would be interested and I told him I thought Mr Irvin would be. Mr Irvin had replaced me when I was moved to the Valve Lab. He had a motorcycle, a Scott Flying Squirrel, a real hot job. I used to sit with him and some of the other draughtsmen from the Research Office in the canteen. I said I would sound him out. I did so at the earliest opportunity and he agreed to join us if Mr Taylor agreed. Mr Chadwick said that Mr Taylor would have to agree.

A technical clerk, Mr Coulthard, was recruited and a lady tracer called Mrs Lumby also started along with Mr Irvin and the office was established in February 1938. Mr Chadwick was really in charge but he left quite a lot of the allocation as to where effort was required to me. The main job was to order all the parts and spares for the 68 cubicles and 34 control desks which were to be built in the West Works. Mr Coulthard said he knew how to use a typewriter so one was obtained for him. I set Mr Irvin on the job of modifying some of the drawings and showed him how the specification sheets had to be written. He now only had to write one copy since Mr Coulthard would type them out. The new tracer set to work tracing some of the general arrangement drawings on to tracing cloth.

The door to the new office was kept locked all the time and only people who were working on the Chain Home Project were allowed in. Mr Irvin was given a key and another key was put in our office in the Valve Lab. A progress chaser from the Research Department Workshop was allocated to work for the Valve Lab and the Research Workshop but he had to give the Valve Lab work priority. His job was to try to control and speed up the flow of material and parts to the various workshops in the factory which were producing parts and making sub-assemblies. If there was any hold-up in the flow of parts he was to report to Dr Dodds or me. Since Dr Dodds was away a lot, or at meetings, or supervising testing, I had to sort out any problems.

The progress chaser's name was Mr Thornton and he had some relations in north Lancashire who had chicken farms. About once a month he would go up north and return on Monday with bags of hen droppings which he sold for 2s. 6d. (12½p each). I still lived in Lower Broughton, Salford, and did not have a garden, but Mr Chadwick, Dr Dodds and Mr Ludlow were regular customers. Mr Thornton's efforts caused a marked improvement on the production and assembly of the CH transmitters and control desks which were now top priority in the whole factory.

By March 1938 further trials using a single aeroplane as the target had been completed at Bawdsey. It was now possible to locate the aeroplane up to 80 miles from the receiver aerials. By sticking a piece of paper across the centre line of the receiver cathode-ray tube and marking the position of the returned signal with the actual positions the pilot gave for the aeroplane, the distance across the 9 inches (23 cm) could be calibrated in miles. The speed of the aeroplane could also be estimated. The position of the aeroplane, whether north, south, east or west of the receiver aerials, was not known nor the height of the aeroplane. Further trials were required when more stations in the Chain Home System were in operation.

Met-Vick was only to supply the transmitters and transmitter control desks – the aerials, supporting masts, aerial feeder cables and receivers were supplied by other companies. We did not know whether we were doing better or worse with regards to deliveries than these other companies but I suspect Dr Dodds would have made it quite clear if we were doing worse.

Dr Dodds was very fond of tea – he did not like coffee and nor did I. In the other offices in the factory they had tea/coffee ladies who brought drinks around in the morning and afternoon. Because of the secrecy of the Valve Lab we could not allow entry to tea ladies, so we had to make our own. We all contributed to the cost of the tea, coffee and sugar and Dr Dodds or Mr Chadwick brought some milk with them each morning. Anyone of us could make the tea but anyone wanting coffee had to make their own and since Dr Dodds and I always had tea, I used to make it quite often. On a number of occasions when I went to make the tea Dr Dodds was there fiddling with what looked like another electric kettle and another teapot. There were also some switches and wiring mounted on a stand. The kettle was sealed and had a pipe coming out of the spout and into the teapot and both it and the teapot seemed to be balanced about the middle of the stand. I never saw the contraption working but I later found out it was an automatic tea-maker. I believe Dr Dodds patented the idea in 1939.

In April 1938 we shipped the equipment for two more Chain Home stations to Swingate, Dover, in Kent and Canewdon, Southend-on-Sea, in Essex. This completed the first orders for five Chain Home stations.

Chapter 7

Chain Home Low and
Coastal Defence Transmitters

IN AUGUST 1938 a much larger trial of the Chain Home System was mounted using a number of aeroplanes coming in over the North Sea towards the Thames Estuary. This was a combined operation using the Army, the Navy, the Coastal Defence and any other defence organisation available at that time. It lasted for a number of days and was co-ordinated from a temporary control centre. All information from the five working Chain Home stations was gathered and assessed. The aeroplanes were located at varying distances from the receiver aerials. From these distances the positions of the aeroplanes, whether north, south, east or west of a Chain Home station, could be quickly calculated and an estimate of the height at which the aeroplanes were flying could be made. This was a great advance since when just one Chain Home station was used the range was the only information which could be established with certainty. (See figure 10.)

Each Chain Home station had four wavelengths to choose from. The reason for this was to combat interference or jamming on any one frequency. The changeover system was tried out on all stations and worked very well. During the combined operation some aeroplanes came in low at about 500 feet above the sea and were not detected by the Chain Home system. The station operators remained unaware of these low-flying aeroplanes until they passed

FIGURE 10

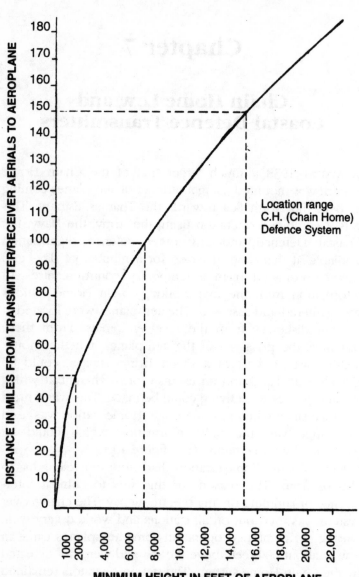

Location range
C.H. (Chain Home)
Defence System

over the Chain Home site. This was very disconcerting. When the defence aeroplanes took off from airfields inland they were also located by the Chain Home stations although they were behind the stations over land and not over the North Sea. This was even more disconcerting. When the results from the combined operations had been digested, a decision was made to have a control centre where all information from all the Chain Home stations and other defence systems could be analysed and the appropriate measures taken for defence purposes. A system to locate the low-flying aeroplanes was required as well as some method of knowing whether aeroplanes were friend or foe.

Dr Dodds and Mr Ludlow had been involved in designing a new modulator oscillator for the Chain Home transmitters. This was probably as a result of the earlier trials carried out at Bawdsey. Both of them had been in our office asking Mr Chadwick and myself about how much room was available in the cubicle for the new modulator oscillator. It appeared to take up more room than the existing unit. To test the new radio equipment Mr Ludlow had made a tone generator which could generate audio tones from about 30 cycles per second to 9,000 cycles per second. By turning the knob of a variable condenser which altered the frequency of the tone being generated, any tone could be heard within the range 30 to 9,000 cycles per second. By a series of quick and slow turns of the knob Mr Ludlow was able to play tunes with the tone generator. He always finished with the National Anthem – about the only tune I could recognise.

In November 1938 Mr Chamberlain, the Prime Minister, went to Munich in Germany to see Herr Hitler to try and get a peace agreement and came back with a piece of paper which he said had secured peace in our time. Irrespective of how Mr Chamberlain's performance has gone down in history, he gained a vital 18 months to enable us to get the Chain Home defence system working before it was really tested during the Battle of Britain in 1940.

Mr Chadwick and I were still working in the Valve Lab although there was plenty of room in the new office. The technical assistant who had left the company early in 1938 had not been replaced, but two new technical assistants were being trained in the Valve Lab extension so that they would get to know the operation of the transmitters and control desks and also be able to carry out any testing required. Mr Ludlow was put in charge of these two.

Since Mr Chamberlain's visit to Munich there was a marked change in the mood of the people as a whole. Most people thought that war with Germany was inevitable and just a matter of time. There was to be an increase in the production of arms and a number of 'shadow' factories were being built for the purpose of producing modern armaments. A new building was to be erected near the West Works just for the assembly of the Chain Home transmitters and control desks and for any other radio location equipment which we might be asked to produce. Dr Dodds told us that we would be getting orders for at least two more different types of transmitters but he did not know how many were required of each type until we got a firm order. There were also to be modifications to the Chain Home transmitters we were now producing to improve their performance. Those already delivered would be modified on site. It was now Christmas 1938 and Dr Dodds suggested that we arrange a party in the new drawing office on Christmas Eve. We should invite Mr Thornton since he was now to be working solely for the Valve Lab. Dr Dodds would provide all the food and drink.

In January 1939 I sold my Matchless motorcycle for £3. 10s. 0d. (£3.50), making a profit of £1 on my buying price. I bought a 350cc Levis (1936 model) for £15. It had an overhead valve engine, a foot change gear box and upswept twin exhaust pipes with polished copper tail pipes – it looked a real modern motorcycle. About the same time Mr Irvin sold his Scott Flying Squirrel and bought a brand new Vincent HRD 500cc with an overhead valve and a sprung rear frame. This was a very fast motorcycle. I could get about 75 mph out of my Levis but he could get about 95 mph out of his Vincent HRD.

In early February 1939 a vehicle which looked like an army trailer appeared and was parked on some waste ground near the Valve Lab. Dr Dodds said that the trailer was to house a transmitter for radio location purposes for Coastal Defence (CD) or any other equipment. It was here so we could assess the amount of space available for the equipment we were to install inside. We were also to manufacture another transmitter which was very similar to the one being installed in the trailer, but which would be sited either on or near the Chain Home site and was to be called Chain Home Low (CHL) transmitter. The basic design for these two new radio location (radar) sets had been done at Bawdsey and they now had some experimental sets operating on a 1.5 metre wavelength, 200 megacycles.

Since we had delivered the first Chain Home transmitter to Bawdsey, Mr Watson-Watt and his team had been carrying out numerous trials and had come to the conclusion that shorter wavelengths would give better results so long as enough power could be produced in the transmitter aerials to get a return signal which could be amplified. A considerable amount of effort was being made to produce glass or silica valves which would oscillate at 300 megacycles or more and generate about 1,000 to 2,000 watts. The valves had to be very compact to generate the high frequencies because it was thought the speed at which the electrons inside the valve were moving was becoming the critical design feature. The smaller the valve the higher the frequency generated, but it was more difficult to dissipate the heat which was developed in the valve. Connections to the filament, cathode, anode and control grid were made of molybdenum which would withstand the heat and not oxidise, but where these connections came through the glass or silica tube, there was a tendency for the glass or silica to melt. Some new insulating materials had been introduced for use on high frequencies as well as cables with centre copper conductors, a new type of insulation material, and a tinned copper mesh screen with insulation overall which could be used for feeders, connections inside the cubicles or connections between cubicles carrying high frequency currents. It had been discovered that on high frequency transmitters and receivers, it was important to use only one earthing point on a chassis or cubicle and not use the chassis or cubicle for earth return as was then the common practice on normal broadcast transmitters and receivers. By earthing at only one point high frequency loops in the chassis could be avoided. With new valves, new insulation materials and new cables being developed, this was the state of the technology in 1939.

Dr Dodds and Mr Ludlow had decided that in view of the amount of drawing effort required for the new orders for 30 Chain Home Low (CHL) and 30 Coastal Defence (CD) transmitters, Mr Chadwick and Mr Irvin in the new office would make the necessary drawings, specifications and order all materials required. Mr Brown (that's me) would stay on the Chain Home transmitters and control desks which were now being produced. There were to be some modifications required on the Chain Home transmitters but Dr Dodds had not decided what was to be altered. Mr Ludlow was to be responsible for the design of the two new transmitters we were to produce. Mr Chadwick was not very happy with this new arrangement as he would have to spend more time in the new office. I had noticed that when Mr Ludlow came into our office and asked questions or wanted any drawings or specifications doing, Mr Chadwick always sought to involve me and I invariably ended up doing what Mr Ludlow required.

In March 1939 Dr Dodds came into the Valve Lab with a gentleman I had not seen before. I knew he must be very important because no one came into the Valve Lab except those who had keys to the doors. Dr Dodds showed him around the Valve Lab extension where the Chain Home transmitters were being assembled and then brought him back into the Valve Lab. Mr Ludlow had got Dr Dodds' 'All Talking, All Singing Arc' going and it was playing one of the classical records. When Mr Ludlow turned up the volume the door on the wire cage behind which the arc was burning began to rattle and anything loose jumped about. After this Mr Ludlow gave a rendering with his new signal generator and finished with the National Anthem, Whilst Mr Ludlow's performance was improving the 'All Talking, All Singing Arc' was the winner.

Dr Dodds called Mr Chadwick and me to meet the gentleman and said this is Mr Watson-Watt. Dr Dodds told him that we had designed the Chain Home transmitters and control desks. I thought he was stretching things a bit as we had drawn it from Dr Dodds' and Mr Ludlow's sketches. But this was how Dr Dodds was – he generously gave everybody else the credit and sometimes they did not deserve it. Even Mr Watson-Watt, the inventor of radio direction finding, was not allowed into the office belonging to Mr Chadwick and me; it was too secret even for him.

A few days after Mr Watson-Watt's visit, Dr Dodds came into our office with Mr Ludlow – this usually meant something had gone wrong but this time we were to get a bit of praise. Now that there were more Chain Home stations working, the information they were collecting during the trials was being directed to one control centre whose purpose was to analyse it and alert the defence forces required. To this end the Chain Home stations were operating better than was expected even at night and in bad weather.

Chapter 8

September 1939 – War

SINCE MR WATSON-WATT'S group had limited resources it had been decided that the RAF would become responsible for the manning, maintenance and security of all Chain Home sites. A man from the Royal Aeronautical Establishment at Farnborough would be coming to arrange for their reference system to be adopted on all the parts of the Chain Home transmitters and control desks. This was to enable spare parts to be specified by using reference numbers and also to enable instructions to be written for the operation and maintenance of all the equipment in use.

An office near our new drawing office was arranged for the man from the RAE who arrived in early May 1939. Dr Dodds introduced Mr Chadwick and me to him and explained that he had to deal with us for any information he required. If he wanted to see any of the items being used on the Chain Home transmitters and control desks one of us would take him along and show him, but he was not to walk about the factory unescorted. When Dr Dodds had gone Mr Chadwick asked the man from the RAE if he could explain what his job entailed. He was about 50 years old. He explained that he was a scheduling clerk and that he was here to give every item, every sub-assembly and every assembly a reference number. Every sub-assembly had to have a 'consisting of' schedule with all the item reference numbers shown and every assembly had to have a 'consisting of' schedule with all the sub-assembly references shown. This was a major job. I got the impression he was expecting to deal with a three-valve set similar to the one I had made in 1931. He would indeed be dealing with a three-valve set, but it was a little different. When he had finished his explanation Mr Chadwick gave him our telephone number and told him that I would help him, I think Mr Chadwick realised just how much help he would require and did not want any part of it.

I went along to the new drawing office and sorted out some of the office copy blueprints to get the RAE man going and asked Mr Coulthard, the technical clerk, to obtain a full set of blueprints for him. When I handed over the blueprints I made it quite clear that they must not be taken out of his office, nor taken back to the RAE when he returned. He had been fixed up with a place to stay in Stretford.

After a few days I went along to his office to see how he was progressing, and by the questions he was asking I realised he had no technical knowledge of the working of transmitters and very little understanding of the drawings I had given him. He had made a list as to why he could not proceed and said he would have to return to the RAE at Farnborough for consultation.

About the middle of May he returned to say that the requirements had been changed. He wanted only one item on each drawing with one drawing number to which he would give an RAE reference number. This would be recorded with a brief title. The sub-assemblies were to be treated in a similar fashion, with one drawing number with all items shown as RAE reference numbers and a reference number for each sub-assembly and also a 'consisting of' schedule for each sub-assembly. The assembly drawings had to be done in a similar manner. At this stage I told the RAE clerk I would have to see Dr Dodds since we could not make any progress until every item had its own drawing sheet and drawing number. When I told Dr Dodds what was required he did not believe what he was hearing and at a fast walking pace we were back in the RAE clerk's office. Dr Dodds pointed out that he had not been told each item had to have its own drawing number and mentioned the names of some of the people he had been dealing with. The RAE clerk said that they were from another section and might not know how the 'consisting of' system had to be applied. Dr Dodds then told him that he wanted a written statement from the head of the clerk's section as to how he wanted the 'consisting of' system to be applied to the Chain Home transmitters and control desks and any other equipment we were likely to make. It was decided that we would do nothing until we heard from the RAE and the RAE clerk returned to Farnborough with Dr Dodds' request.

Early in June 1939 we received a letter from RAE Farnborough confirming all that the RAE clerk had said. In addition they wanted the drawings for the Chain Home Low transmitters to be done in a similar manner. When Mr Chadwick had read the letter he said we would like two more tracers as soon as possible. It was also decided that the Coastal Defence transmitters and any other work we might be allocated would be done on the same basis: one item shown on one drawing with one drawing number for each item.

We very quickly obtained the services of two more lady tracers from other drawing offices in the factory and arranged for them to go into the new drawing office. When Mr Chadwick and I had been moved to the Valve Lab in 1936 it was intended that we should stay for about six months – we had now been there for three years and the demand for the equipment we were designing seemed to be growing.

By the middle of 1939 there were three tracers (ladies), three draughtsmen and one technical clerk as well as Dr Dodds and Mr Ludlow all engaged in producing drawings and specifications for the Chain Home, Chain Home Low and Coastal Defence transmitters. The Chain Home transmitters and control desks were being built in the Valve Lab extension, in the area partitioned off in the West Works switchgear building and in the new building near to it.

I got all three tracers underway by taking blueprints of the detail drawings and cutting them up so that each item could be traced onto tracing cloth and given its own drawing number. All drawings in the factory were done on 'A' size sheets (40 in. x 30 in.), 'B' size sheets (30 in. x 20 in.), 'C' size sheets (20 in. x 15 in.), 'D' size sheets (15 in. x 10 in.). All the drawings done so far had been on 'A' size or 'B' size sheets, but now most of the drawings would be 'C' size and (mostly) 'D' size sheets. We would requite a lot more drawing numbers. We had also decided to keep a set of drawing numbers in sequence for the Chain Home drawings, a set of numbers for the Chain Home Low drawings and a set for the Coastal Defence transmitters. A further 300 drawing numbers were allocated to us by central file and our technical clerk started two more drawing number record books, one for Chain Home Low drawings and the other for Coastal Defence drawings. Mr Irvin had started on the drawings for Chain Home Low transmitters. Mr Chadwick was doing the drawings for the Coastal Defence transmitters and, with some help from our technical clerk, I was cutting up blueprints to keep the tracers busy.

In July 1939 Mr Irvin was killed in a motorcycling accident. Two more draughtsmen joined us from other drawing offices in the factory and Mr Chadwick asked me to show them what was required and to interpret for them Mr Ludlow's sketches for the Chain Home Low transmitters. These were different from the Chain Home transmitters: they were designed to operate on a wavelength of 1.5 metres – a frequency of 200 megacycles with a maximum output of about 10 kW. On account of the shorter wavelength the transmitted beam could be made very narrow and directional. This was not possible with the Chain Home transmitter working on 10 – 15 metres where the transmitted wave was more like a floodlight – it covered a wide angle but was not very directional. The wave tended to balloon up into the sky and consequently was very good for locating high-flying aircraft but could not locate low-flying ones. By using the 1.5 metre wavelength and, as the technology advanced, even shorter wavelengths, the development of radio direction finding was advancing out of all proportion to anything envisaged in 1936.

The Chain Home Low transmitter was arranged in a sheet steel cubicle with rolled steel angle framework and was about 5 feet high, 5 feet wide and 3 feet deep. It was divided up into three chambers. One was for the power unit which was a 15 kVA transformer with glass rectifying valves and smoothing condensers providing the voltages required for the filaments, anodes and control grids for the transmitting glass valves. There was a chamber for the modulating and transmitting valves and a chamber for the aerial coupling unit. It was designed to operate only on a 1.5 metre wavelength. It did not have any wavechange system as had been provided in the Chain Home transmitter and was modulated by the 50 cycled national grid supply similar to the Chain Home transmitters. On the front panel was mounted a six-inch cathode-ray tube for monitoring the wave form being transmitted, together with ammeters, voltmeters, variable resistances and switches for controlling the valves.

In August 1939 the RAE clerk returned to see if we had made any progress in creating drawings of each item for the Chain Home transmitters and control desks. I showed him some of the drawings the tracers had produced; he seemed quite happy and said he could now make a start on allocating reference numbers so that the tracers could include them. Difficulty arose when he thought an item we had used on the Chain Home transmitters was the same as one already used elsewhere on other RAE equipment. He didn't want two different numbers for the same piece of equipment so any doubtful item was laid to one side to be dealt with later.

When the RAE clerk had said that what they wanted was one item on its own drawing I was of the opinion that this was a waste of time and effort, but it proved to be a good thing in the long run since the drawings would have been traced anyhow. We also found it advantageous to have one item on its own drawing number when we specified parts to be made in the factory on our own specification sheets. Meanwhile Mr Chadwick continued working on the drawings for the coastal defence mobile transmitter. While it was very similar to the Chain Home Low transmitter, it was made in three sheet steel cubicles with rolled steel angle framework. Each of these was about 5 feet high, 2 feet wide and two feet deep. They were shipped as separate units and assembled in the trailers by some other manufacturer.

In September 1939 we were at war with Germany. Our work on the Chain Home stations had reached a point where we had supplied transmitters and control desks for about 20 stations from Ventnor in the Isle of Wight to Hillhead in Scotland. Some of the stations may not have been complete but most were. The only Chain Home Low and Coastal Defence transmitters working were the experimental ones in the Bawdsey area. I don't know whether the control centre – where all the information was received and where the necessary defence was activated – was working or not in September 1939.

The RAE clerk was still coming and going and he said the people at Farnborough were quite pleased with the progress we had made. I was of the opinion that he had done the easy ones so far and had put the difficult ones to one side. A difficulty arose when he had to allocate numbers to the valves. Practically all radio valves used at this time had a glass envelope mounted on an eight-pin base and reference numbers prefixed with VR (valve radio) were given for each different type of valve used by all three services. There was therefore a certain amount of standardisation. If a valve required changing all that was required was the VR number and an exact replacement could be obtained by non-technical personnel. All radio manufacturers had their own system of classifying valves and each services VR number covered the different manufacturers' reference numbers for any given type.

The valves used in the Chain Home transmitters were demountable and all the internal and external parts would each have to be given a reference number, and a sub-assembly number for the filament, control grid and suppresser grid assembly. I explained to the RAE clerk how the valves were made. The joints had to be lapped to make them vacuum tight and if they carried spare vacuum pumps, spare valve bases, anodes and ceramic insulators, they would have to be lapped on site which might prove very difficult in the places where the Chain Home stations were sited. The RAE clerk decided he would have to return to Farnborough for consultation.

Dr Dodds had been away for a few days in October 1939 and on his return he said that there were some modifications required on the Chain Home transmitters which would improve their performance. He had some sketches showing what was required. The main modifications concerned the feeders between the two cubicles and there were to be changes to some of the tuning coils and larger condenser bushings in the power supplies to the valve anodes. He also had an outline drawing of a new modulator which was being supplied by the manufacturer of the receiver. I had to ensure it could be fitted in the compartment we had provided in the transmitter cubicle.

Since all the Chain Home transmitters had been shipped and were now on the various sites along the south and east coast, these modifications would have to be carried out on site. The reason for all the modifications was to change the transmitters from a modulated continuous wave class 'B' type of transmission to a pulse-modulated class 'C' type of transmission. The new pulse rates were 25 and 12.5 pulses per second. Pulse length was 15 micro-seconds and this was controlled from the receiver. With the modified arrangement the Chain Home transmitter could deliver a pulsed power of 650 kW which increased the aeroplane detection range to at least 150 miles. This was to be a great advantage in 1940.

I did not change the original drawings and specifications since that would have confused the RAE clerk with his reference numbering. I made new drawings in line with the new system of one item per drawing and ordered all the new parts required. We were aiming to have all the necessary parts and instructions ready and delivered to the sites so that all the Chain Home stations could be modified in March 1940.

Early in December 1939 the RAE clerk returned and said a decision had been made regarding the demountable valves. They were not to have a VR (valve radio) number but were to be given a number just like any other sub-assembly and all the items were to have separate numbers. Parts which were to be lapped together as a unit had a note to that effect added – this was the best we could do.

We had just about finished putting reference numbers on all the detail drawings and about all that remained to be done was to make up a 'consisting of' schedule for the transmitters and one for the control desk. Transmitter type T3026 had been allocated for both cubicles and control desk. I now had a very good idea as to how the reference numbers were to be used together with the drawings in compiling the 'consisting of' schedules, so I suggested to the RAE clerk that it would be better if I wrote them out and he could check them when he returned. Since it was now coming up to Christmas 1939 he agreed and said he would return in January 1940.

The RAF had been sending bomber planes over Germany carrying propaganda leaflets which were scattered far and wide. When they returned the Chain Home stations could locate them but it was impossible to know whether the aircraft they had detected and which could be seen on the cathode-ray tubes were 'ours' or 'theirs'. As the aeroplanes came in over the coast they were greeted with anti-aircraft fire. At this stage in the war the anti-aircraft batteries were not very accurate and just relied on listening devices, so little damage was done to the returning aircraft. Enemy aeroplanes received similar treatment and not much damage was done to them either.

Chapter 9

Friend-Foe Indicators
and Airborne Transmitters

AS A RESULT of this confusion a device had been developed by Mr Watson-Watt's team which became known as an FFI (a Friend-Foe Indicator) and about the middle of December 1939 Dr Dodds was given an order and a specification to produce ten units. Mr Ludlow was to be responsible for the design and I was to produce the necessary drawings and parts specifications. Up to now all the radio equipment we had produced was quite large – this was quite small. It was about 12 in. wide, 12 in. high and 18 in. long and was to be as light as possible. Since it was to be airborne the valves had to be restrained to stop them jumping out of the valve holders. All fixing screw threads had to be smeared with 'lok-tight' glue, all soldered connections had to be made fast before soldering and the box containing the whole unit had to be finished in non-reflecting black enamel. Any plug and socket connections had to have a screw-down retaining ring. It had three glass valves with all the components mounted on a small chassis.

When the set was switched on, the tuning condenser was driven continuously to cover for Chain Home stations on 20 – 30 megacycles and 190 – 220 megacycles. If none of these frequencies was detected the set would continue to search until it was switched off. If any one of these frequencies was detected the set would start to transmit on all frequencies and this transmission would appear as a pulsing spot

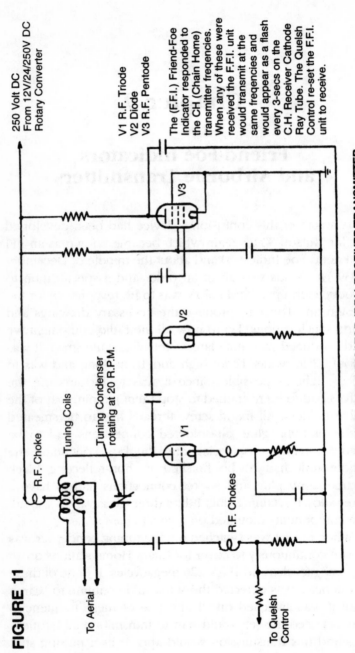

FIGURE 11

R.F. Choke

Tuning Coils

Tuning Condenser rotating at 20 R.P.M.

To Aerial

V1

R.F. Chokes

V2

V3

To Quelsh Control

250 Volt DC
From 12V/24/250V DC
Rotary Converter

V1 R.F. Triode
V2 Diode
V3 R.F. Pentode

The (F.F.I.) Friend-Foe Indicator responded to the CH (Chain Home) transmitter frequencies. When any of these were received the F.F.I. unit would transmit at the same freqencies and would appear as a flash every 3-secs on the C.H. Receiver Cathode Ray Tube. The Quelsh Control re-set the F.F.I. unit to receive.

DIAGRAM — FRIEND-FOE INDICATOR (F.F.I.) RECEIVER/TRANSMITTER

of light on the Chain Home receiver cathode-ray tube at the appropriate distance in miles from the Chain Home receiver aerials. This indicated the aeroplane they had detected was a friend since the spot of light on the cathode-ray tube due to any enemy aeroplane without an FFI was a steady spot of light and did not pulse.

All aeroplanes were fitted with either a 12 volt DC battery electrical system similar to that of a motor car or a 24 volt battery electrical system similar to that of a lorry. Since the radio valves required about 300 volts DC for the valve anodes, a rotary converter had to be provided for each Friend-Foe Indicator to increase the voltage from 12 or 24 volts DC to 300 volts DC. The box containing all the equipment was made of $1/16$ in. thick aluminium sheet and was black stove enamel shrivel finish which was non-reflecting. There was a switch on the aeroplane control panel for switching the Friend-Foe Indicator on or off. The pilot of the aeroplane was told to switch it on when over friendly areas and when approaching the British coast. Over enemy territory the switch had to be off. This was probably the first mysterious black box. (See figure 11.)

We made the first Friend-Foe Indicators as prototypes in the Valve Lab extension and they proved to be successful. Every Allied aeroplane was fitted with one so there must have been quite a lot made in our factory and by other firms during the war. As the number and types of radio location sets increased, the original Friend-Foe Indicators were modified and new types of FFI unit were developed to respond to the shorter wavelengths being used.

It was now Christmas 1939 but we did not have a party because we had to work during the holiday and have the time off later. Since most of the radio location sets were being made in other parts of the factory, the Valve Lab extension was being used for making prototypes and developing new equipment. Dr Dodds and Mr Ludlow had installed quite a lot of test gear for testing radio circuits at different stages of manufacture and were also designing

and making jigs and fixtures for wiring harnesses. When
these had been tried out they were moved to other parts of
the factory where the radio location sets were being
assembled. Test gear was also being installed for testing
parts coming from outside manufacturers such as valves,
cathode-ray tubes, condensers and resistors, both fixed and
variable.

Condensers and resistors were only accurate to plus or
minus 10 per cent of the stated values, which was good
enough for most purposes at the time, but greater accuracy
was required for some of the circuits now being developed.
So all condensers and resistors were tested and divided into
three categories: those minus 10 per cent of the
specification, those with plus or minus 2 per cent of the
stated value and those 10 per cent above the stated value.
By selective assembly better and more accurate circuits
were produced.

The orders for 30 Chain Home Low and 30 Coastal
Defence transmitters were almost complete and we had
shipped most of the equipment. In January 1940 further
orders for another 80 Chain Home Low transmitters and 60
Coastal Defence transmitters were received. All the sets
were to be modified so that they would be modulated in a
similar fashion to the chain Home transmitters with a pulse
rate as a multiple of the 50 cycles per second national grid
frequency. The Coastal Defence power was supplied from
mobile diesel generators and these were not synchronised
to the national grid frequency. The modification to the
system of pulse modulation for the Chain Home Low transmitters improved their performance and aeroplanes upto
seventy miles away could be located. Coastal Defence
transmitters could locate ships and low-flying aeroplanes
up to thirty miles away depending on how high the
transmitting and receiving aerials were above sea level.
(See figures 12 and 13.)

We also received an order for 30 airborne transmitters
and Dr Dodds said he would be giving us details of what

was required. These were the first operational airborne radio location sets to be made and they were intended to counter the nightly intrusions of German aircraft by locating them and so that they could be shot down. They were also the first radio location sets which were made for attacking – all the previous ones such as the Chain Home, Chain

FIGURE 12

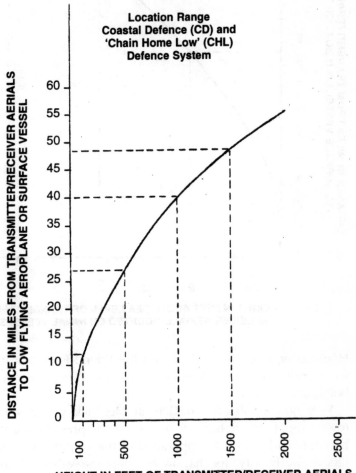

Location Range
Coastal Defence (CD) and
'Chain Home Low' (CHL)
Defence System

FIGURE 13

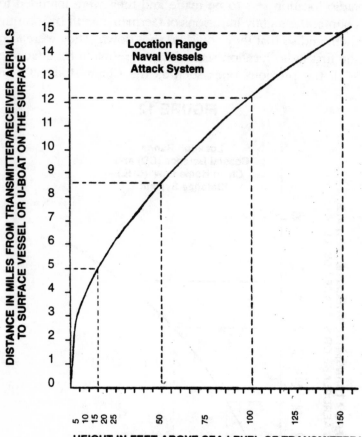

DISTANCE IN MILES FROM TRANSMITTER/RECEIVER AERIALS TO SURFACE VESSEL OR U-BOAT ON THE SURFACE

Location Range
Naval Vessels
Attack System

**HEIGHT IN FEET ABOVE SEA LEVEL OF TRANSMITTER/
RECEIVER AERIALS MOUNTED ON NAVAL VESSELS**

Home Low, Coastal Defence and the Friend-Foe Indicator were primarily for defence purposes. The tide was beginning to turn.

Very near to the South Gate at Met-Vick was a water tower about 200 ft high. It was used for pressurising the hydrants and also in the operation of the various hydraulic lifts in the factory. Its shadow fell over the Valve Lab and most of the Research Department when the sun was shining. The management at Met-Vick, probably in consultation with some military expert, decided that the water tower must come down as it could be a location point for German bombers. So in January 1940 it was very quickly reduced from 200 ft to about 20 ft and as a result a large consignment of scrap steel went to the war effort. I do not know its capacity but it was probably a few thousand gallons. At 200 feet high the water pressure at ground level would be about 90 pounds per square inch. All that water and water pressure would have been very useful when the factory was bombed in December 1940. The fire pumps were not strong enough to pump the water high enough to reach the burning roofs which were about 80 ft high. With 90 pounds per square inch pressure available at ground level the pumps would have thrown the water jets much higher than 80 ft.

Towards the end of January 1940 the RAE clerk returned to see how I was getting on with the 'consisting of' schedules. They were the longest and most detailed documents I had ever written. I spent some time going through the various schedules with the RAE clerk and making some modifications which he requested and he finally agreed that they were to his satisfaction. I had kept Mr Chadwick in touch with what I was writing and he had read most of it by the time I had finished. Dr Dodds and Mr Ludlow had not seen any of the documents so I told the RAE clerk that we should arrange for them to be given the opportunity to see them if they wished.

When I told Dr Dodds that we had finished the 'consisting of' schedules he seemed quite relieved and asked me if the RAE clerk had agreed they were complete. When I told him the RAE clerk was quite prepared to accept all the documents and take them back to Farnborough, he was even more pleased. We collected Mr Ludlow and all three of us went to the RAE clerk's office who confirmed what I had already told Dr Dodds. Dr Dodds congratulated the RAE clerk and me for making such a good job of the 'consisting of' schedules. He also pointed out to the RAE clerk that any schedules required for other equipment would have to be done at Farnborough by RAE personnel as we now had too much design work to do and could not get involved in schedules. I think it was the biggest consignment of 'consisting of' schedules for radio equipment that the RAE had ever made.

The RAE clerk returned to Farnborough with all the documents and we did not get involved in any other 'consisting of' schedules for equipment we designed and produced after 1940. Mr Chadwick had made a start on the 30 airborne transmitters we were to produce; now that we had a firm order we could start making drawings and order components. As I had finished my work for the RAE clerk, Mr Chadwick asked me to have a look at the sketches and specifications Dr Dodds had given to him. We were to provide the power supplies for one transmitter on 1.5 metre wavelength (200 megacycles), two receivers on 1.5 metres and two 6 in. cathode-ray tubes which came to a total of about 1500 watts for each aeroplane being fitted with the airborne radio location sets. The aerials and aerial feeders were to be supplied by other sources as were the receivers and cathode-ray tubes. We were also to supply the transmitters and the generators for the power supplies.

These night-fighter aeroplanes had to be much larger than the fighter aeroplanes used in daylight. They would have a pilot and an operator for the radio location equipment. On the front of the plane would have to be mounted three half-wave di-pole aerials: one for the transmitted pulses, one vertically polarised for up-down reference and one horizontally polarised for left-right reference to the target. Space had also to be provided inside the body of the aeroplane for the power generator, the transmitter, two receivers, two cathode-ray tubes and all the additional wiring required. All this extra equipment would affect the speed and performance of the aeroplane.

The theory at this stage of the development of radio location was that at night and in very cloudy conditions any enemy aeroplanes approaching the coast would be located first by the Chain Home sets at about 100 to 150 miles from the coast and would be tracked by the Chain Home Low sets and Coastal Defence sets. The new night-fighter aeroplanes with their new radio locators would be alerted and 'talked' into positions so that they could intercept the enemy aeroplanes by means of the normal radio communication system which operated on a 50 metre wavelength (6 megacycles). On the cathode-ray tubes at the Chain Home stations the enemy aeroplanes would appear as moving spots of light, whereas the night-fighters would appear a pulsing spots of light (so long as the Friend-Foe indicators were switched on and working).

The new radio locators which were to be installed in the night-fighter aeroplanes had a maximum range of about 25,000 ft and a minimum range of about 1,000 ft. When an enemy aeroplane was located by the night-fighter the radio locator operator could instruct the pilot to go up, down, left or right as a result of what he was seeing on the two cathode-ray tubes. As the distance between the two aeroplanes was reduced it was hoped that the night-fighter pilot would see the flames from the exhaust of the enemy aeroplane. This was the theory, but in January 1940 the drawings for the airborne radio locators were still in the process of being made.

As I explained earlier, the electrical power supply even on the larger aeroplanes was only 24 volts DC, which was derived from batteries which were charged by a DC generator driven by one of the engines. To be able to provide all the different voltages for the valve filaments, cathodes, grids and anodes in the transmitter and receivers and also those for the filaments, external deflection coils, internal deflection plates and an electron beam accelerating gun in the cathode-ray tubes, would require an alternating-current generator to be provided for each night-fighter aeroplane.

Since the alternating-current generator was to be used

only for powering the radio equipment, the frequency could be raised to 500 cycles per second instead of the 50 cycles per second of the national grid. By raising the frequency to 500 cycles per second the size and weight of transformers, chokes and condensers were reduced so that the transmitter and the two receivers were more compact and lighter than they would have been if we had used 50 cycles per second generators. This reduction in size and weight was a big advantage when more electronic devices were added to aeroplanes as World War II progressed. Later, higher frequency generators were used in aeroplanes – our 500 cycles per second was the beginning of the process.

The airborne transmitters were a smaller version of the Coastal Defence sets we had made. The oscillator and modulation valves were made of glass and measured about 8 in. high and 3 in. diameter; they rated at 1,000 watts maximum. The valves were modulated by using thyratons with timing circuits and sharp cut-off circuits, similar to those on the Chain Home and Chain Home Low sets, but the number of pulses per second was increased to 3,000 with a pulse length of 10 micro seconds.

There were three sheet aluminium boxes measuring about 18 in. high x 18 in. wide x 15 in. deep. One housed the transformers, rectifiers, condensers and terminals to form the power supply unit, another the oscillator, modulation valves and thyratron timing equipment and another was for the coupling coils and aerial connections. The boxes were arranged to be counted vertically or horizontally according to the space available. There was also a small centrifugal air-blower provided to dissipate the heat generated by the valves. The air blower was powered by the 24 volt DC aeroplane electrical system. These airborne transmitters were called Air Intercept (AI) transmitters and by the time we made the drawings and got them into production they became Air Intercept (AI) Mark IV. The drawings and the issuing of manufacturing instructions were all carried out in the new drawing office.

Up to this point in the war most of the enemy attacks had been against ships in the shipping lanes along the east and south coasts. The enemy had also been laying mines in these shipping lanes and in estuaries leading to docking areas and ports. The Germans had developed magnetic and acoustic mines which had been sinking a number of ships around the coasts. The mines were dropped by low-flying aeroplanes but we now had a number of Chain Home Low and Coastal Defence stations which could locate and track the enemy planes. By following the tracks the enemy planes had taken, the minesweepers had a better chance of exploding the mines. The minesweepers had to have wooden hulls and were fitted with de-gaussing equipment which neutralised any magnetic effect created by the engine propeller and propeller shaft. Towed behind the minesweeper was a barge with cables carrying pulsed DC currents to give a magnetic field. On board it was also a device which flapped on the water and made a noise like a ship's propeller. The engine speed of the minesweeper was kept as low as possible to keep the engine noise and propeller noise down to a minimum. All ships which used the shipping lanes along the south and east coasts were fitted with de-gaussing equipment as it became available. Through a combination of the Chain Home Low and Coastal Defence stations and the modified mine-sweepers, the threat posed by magnetic and Coastal Defence stations was overcome, but it took time. The shipping lanes along the south coast and into the Thames estuary became known as 'Bomb Alley'.

Chapter 10

German Air Force
Tests the Chain Home System

ALL THE PARTS for the modifications to the Chain Home transmitters had now been manufactured and delivered to all the Chain Home sites. Mr Ludlow said that he and the two technical assistants from the Valve Lab extension would be visiting one of the Chain Home sites on the east coast to check if there were any difficulties with the parts supplied and to carry out some tests. He asked me to accompany them with a set of the modification drawings and any other drawings I thought might be useful. We would be going early in March 1940 and it was very important to ensure there were no snags.

Early in March 1940 the four of us set off in Mr Ludlow's car with some testing equipment and some parts which he thought might come in useful. I also had my set of blueprints. Once we set off Mr Ludlow said we were going to the Chain Home station at Danby Beacon. We were to stay at the Sandsend Hotel, Sandsend. As we approached Danby Beacon the next morning I could see the four 350 ft masts and as we came near to the Chain Home station there were RAF personnel guarding the site.

The building which housed the transmitters was partly buried in the side of a hill and looked a very strong structure. The door was very substantial and had metal cladding and there was a sandbagged passage leading to it. As we went through the door into the building I noticed that the transmitters and control desks were arranged as we had them arranged in the Valve Lab. Their positions were decided by the feeders from the top of the cubicles to the aerial coupling unit.

When a check had been made that all the parts required had been delivered and were available and it had been decided which of the two transmitters was to be modified first, Mr Ludlow and the two technical assistants started to carry out the removal of the feeders between the two cubicles. The existing feeders were to be re-used with additional lengths fitted to them and since there were four wavelengths to deal with, the feeders had to cross over each other. It was a bit complicated but I was fairly certain that my drawings were correct and the parts would fit.

We were allowed a short lunch break but nonetheless Mr Ludlow seemed quite pleased with the progress we had made so far and hoped to be able to complete the modifications by about 7 p.m. that day so that the modified transmitter could be put back on the air overnight. The other transmitter could be altered the next day. The transmitters were as we had designed and made them but some additional labels had been fitted by the operators to make it easier for them. Labels had also been added to the control desk.

By the time the new modulator unit, the larger condenser bushings, the new tuning coils, new copper tube and the internal connections had been installed and the feeders between the cubicles and to the aerial coupling unit had been refitted, it was about 6 p.m. Mr Ludlow and the two technical assistants carried out some tests to ensure no obvious mistakes had been made and by about 7 p.m. we were on our way back to Sandsend. As Mr Ludlow pointed out, the real test would come when the transmitters were put back on the air. He also said he had left the hotel telephone number with the Chain Home station operators, but we were not disturbed and returned the next morning.

As we went through the door of the transmitter building

I could see that the transmitter we had modified was on the air. The indicator lamps, ammeters and voltmeters on the control desk were all working. The operator said everything seemed to be in order; the other transmitter was shut down and could be worked on. I was quite relieved to find that these modifications had gone so smoothly. Mr Ludlow and the technicians had a look around the cubicles and checked the readings showing on the various instruments on both cubicles and the control desk and they seemed quite happy with the overall results.

Mr Ludlow congratulated me on getting the drawings right first time and the two technical assistants on a job well done. There was no one to congratulate Mr Ludlow. The second transmitter was modified in a similar manner to the first and by about 7 p.m. we were back at the Sandsend Hotel. Next morning Mr Ludlow said we would return to Manchester but would call in at Danby Beacon on our way. After about an hour of checking and talking to the operators everybody seemed to be in agreement that the Danby Beacon was working better than it was before the modifications had been made, but the station had to be re-calibrated for the increased range the transmitters were now capable of producing. These modifications had changed the transmitters from the modulated continuous wave class 'B' type to a pulse-modulated class 'C' type of transmission, with a pulsed power of about 650 kilowatts. They had increased the aeroplane detection range to at least 150 miles. This type of pulse modulation was a big break through and was subsequently used on all types of radio locators during World War II.

From the start of World War II in September 1939, the operators at the Chain Home sites and Coastal Defence sites had a lot more aeroplanes to locate and track, both friend and foe, than they had up to September 1939. As a result of increased activity they were beginning to understand in more detail what they were seeing on the cathode-ray tubes and were able to estimate the number of aeroplanes coming towards the coast, their direction, speed and height and whether they were friend or foe. Now that the detection range had increased to at least 150 miles there was about 30 minutes, notice of a possible air raid if the aeroplanes were foe, and the point where the aeroplanes would cross the coast could be determined and also possible targets. This information was fed to a control centre where all the details of the aeroplane locations coming from the various Chain Home and Coastal Defence stations could be assessed so that the appropriate defence forces were alerted. From Ventnor in the Isle of Wight along the south coast to Dover, and along the east coast to the Orkney Islands the 22 Chain Home stations were on watch 24 hours a day, every day.

The whole defence system was still very secret. There had been no mention of it in any newspaper or technical publication and so far the enemy had not shown any interest in the four 350 ft masts spaced about every 30 miles along the south and east coasts. The enemy reconnaissance aeroplane pilots must have seen the masts but did not report them to the people who would know their purpose. We were not sure in April 1940 whether the Germans had developed any radio location system similar to our own and since they had not so far attacked any of the Chain Home sites, it was generally thought that they had not. Our radio location system was purely a defence system. Since September 1939 the Germans had done all the attacking and had not been on the defensive. At this stage of the war they could see no reason for air defence measures since up to this point they had been winning. There had been no attempts by the enemy to interfere with or jam any of the frequencies we were using on the Chain Home and Coastal Defence sites. The Chain Home Low and Coastal Defence transmitter and receiver aerials could be turned through almost 360 degrees, but this operation had to be done by hand and it took a few minutes to go the full circle.

It had been noticed that when the receiver aerials were

pointing in the direction of the aeroplane which had been located, a clearer image was shown on the cathode-ray tube. Furthermore, if the transmitter aerials were pointing in the same direction as the receiver aerials, the image was even better, with less 'clutter' as interference came to be known. As a result of these observations it was decided to power-operate both aerials so that a large area could be swept in about one minute under the control of the operator watching the cathode-ray tube. A more exact location of each aeroplane could also be given.

At this stage of the development of the radio location defence system, friend or foe aeroplanes could be located and tracked coming into or going away from the east and south coasts, but once they had crossed the coast and were over the land they could only be tracked by the noise they made. This was true even in daylight. Some people said they could tell whether an aeroplane was one of theirs or one of ours by the rhythmic beat of the engines, but I was never sure. As a result of the improvement of the Chain Home Low radio locators resulting from the modifications to the transmitter and receiver aerials, it was decided to site radio location sets inland to track all aeroplanes whether friend or foe which were flying over the land. These were to be called Ground Control Intercept (GCI) stations. They were to operate on a 1.5 metre wavelength (200 megacycles) and were similar to the Chain Home Low sets with a different aerial array, and would have a range of about 80 miles. Since the aerials could be turned through one revolution in about one minute, an area of 160 miles' diameter could be swept in that time. By siting these GCI stations in various parts of the country, it was possible to track all aeroplanes as they came in over the coast and then overland. It was hoped that with the advent of the Air Intercept sets we were now making, some of the enemy aeroplanes coming over at night would be located and shot down, but in April 1940 the equipment was not yet available.

In May 1940 the war in Europe started in earnest and in
a matter of weeks the Germans had control of the European
coastline from the north of Norway as far as Spain. By June
1940 they had control of all the French, Belgian, Dutch,
Danish and Norwegian airfields which would be very
useful to them for attacking us. With the fall of France, the
Chain Home defence system west of Ventnor along the
south coast was very exposed and every effort was made
to close this gap by using Chain Home Low and Coastal
Defence sets. Whereas these sets did not have the range of
the Chain Home sets, they were able to give an early
warning of about 15 minutes of friend or foe aeroplanes
coming in over the coast.

The following month there was a real threat of a German
invasion and since quite a lot of our military equipment had
been left in France, there was a general shortage of every
kind of arms. A call went out asking people to form Local
Defence Forces (which became known as the LDV – Local
Defence Volunteers) in every part of the country. I went
along to Cross Lane Barracks in Salford to sign on. It was
chaotic – there were thousands of men of all ages wanting
to get involved. I eventually signed on and was given a
time and date to do some training. I still had my Levis
motorcycle and thought I might get an extra petrol ration.
I went along to the barracks at the stipulated time to find
that there were about 30 men of all ages there. We were
instructed as to what we had to do if we heard the church
bells ringing. This was the agreed signal throughout the
country if an invasion was imminent. We were divided up
into sections of six and each section was taken in turn to a
firing range in the basement. An instructor had a 0.22 in.
cartridge gun and each of us in turn was allowed to fire
three shots at a target. I do not know whether I hit or
missed it – we were never told. We stayed in the same
group and were given a 0.303 in. rifle with no ammunition
and were given the task of guarding the barracks for the
night. Two of us were on guard for two hours while the

others could do what they wanted so long as they didn't leave the barracks. The man I was with had been in Germany in the Army of the Rhine after World War I was over. He showed me how the rifle worked, how to load it and clean it and also told me tales about his experiences in Germany. After about two hours we were relieved and I got down on the floor with a blanket and tried to sleep. At about 6 a.m. we were wakened and someone had made tea. We were told to come back a week later at the same time. As a result of the threat of invasion all road signposts were removed to make it more difficult for the enemy if parachute troops or saboteurs were dropped. The Local Defence Volunteers had to be local men so that they knew their way around without the use of the road signposts.

Towards the end of July 1940 the Germans began to step up their attacks on shipping along the east and south coasts. It was now our turn for the German bombing treatment. So far the German army and air force had carried all before them; all Germans were right behind Hitler and he was winning. There is nothing like a winner in warfare and consequently our leaflet raids had no effect at all on the German people. We were lucky – thanks to Mr Chamberlain's efforts in Munich in 1938, we had had 18 months of respite of which we had made good use.

By early August 1940 there were daily air battles over the south and south-east coasts and quite a number of the German aeroplanes were being shot down. This was the beginning of the real test of the Chain Home defence system which we had started work on in the Valve Lab at Met-Vick back in 1936. This was the most important battle so far from the British viewpoint. If we had lost, the war could have been over for us and probably for a lot of other countries as well. 1940 was a brilliant summer and the Germans started attacking aerodromes and other installations to the south and east of London. There were spectacular aerial combats over Kent and the surrounding countryside. The enemy losses in aeroplanes and men were beginning to increase, and they must have wondered how it was that, irrespective of how they approached the east and south coasts, there were usually RAF fighter aeroplanes in a good position to attack them.

These air battles continued to about the middle of August 1940. About 1,000 German aeroplanes had been shot down and there was speculation in some of our newspapers and also in foreign publications that the British had some sort of secret weapon for destroying aeroplanes. There were all kinds of ideas suggested: a death ray, a ray which penetrated the fuel tanks and set the aeroplane on fire, a device which upset the ignition system of the aeroplane engines and many others. There was no confirmation or rejection of these suggestions by the Ministry of Information and this added to the mystery.

The Chain Home defence system was still working and remained very secret. As a result of the modifications to the Chain Home transmitters in March 1940 when the range was increased to 150 miles, the stations at Dunkirk, Dover and Rye could be used to watch the German aeroplanes forming up over the French airfields in preparation for an attack. An estimate of the number of enemy aeroplanes in each attack could be made as well as an assessment of the point where each flight would cross the coast. With this information available at an early stage of each attack, the defence forces had time to organise the RAF fighter aeroplanes, anti-aircraft batteries and balloon barrages to be ready and waiting for the enemy aeroplanes to appear. The Germans tried some diversionary attacks along the east coast but were confronted by defence forces. From these attacks they must have concluded that the 350 ft high masts spaced about every 30 miles had something to do with their aeroplanes being located so that a defence system was always ready and waiting for them.

About the middle of August 1940, determined attacks were launched against the Chain Home stations along the south coast and those around the Thames estuary. Ventnor was badly damaged and Poling, Pevensey, Rye (Dungeness), Dover (Swingate) and Dunkirk (Foreness) were all damaged to a lesser degree. Although these attacks made a hole in the defence system, the Chain Home Low and Coastal Defence sets were still working, as were some of the Ground Control Intercept stations which had now been installed. Some of the technicians who had worked on the Chain Home transmitter at Met-Vick were sent to help with the repairs to the stations along the south coast.

These attacks continued almost every day until the middle of September 1940, when the Luftwaffe mounted a very large scale and sustained attack on airfields in and around London and there was an outcry that Berlin should be bombed. So far only leaflets had been dropped on Germany. Towards the end of September 1940, the daylight air attacks started to decrease and there were far fewer aeroplanes involved. The Germans had been made to confine themselves to hit and run raids with single aeroplanes. Nonetheless they were still located by the Chain Home defence systems and either shot down or chased back to France. The Luftwaffe had lost about 1,800 aeroplanes and a great number of pilots and navigators. This was their first defeat and with such heavy losses, they realised that it was too dangerous for them to send aeroplanes in daylight over the south and east coasts.

Early in October 1940 the Luftwaffe started night bombing attacks on London and other towns and cities. They had developed a system of radio beams to help their aeroplanes find their targets. A radio transmitter sited on the Channel coast would direct a pulse-modulated radio beam over the target and the pilot would hear the pulse when his aeroplane was in the correct position of the beam. A receiver in the aeroplane would pick up the radio pulses and retransmit them on a different wavelength back to a receiver at the transmitter station. By displaying the outgoing pulse and the received pulse on a cathode-ray tube and measuring the time which elapsed, the distance of the aeroplane from the transmitter could be shown and the operator at the transmitter end could tell the aeroplane pilot where he was relative to the target using the radio communication system. This arrangement for locating targets was very accurate and caused quite a lot of consternation when the beams were discovered. The answer was to jam the transmitted or the re-transmitted beams or better still jam both. But in October 1940 we didn't have the necessary equipment to jam either of the transmissions. When the target was located it was marked with parachute chandeliers and incendiary bombs and the following aeroplanes just bombed the marked area.

The Chain Home system had now been extended along the south coast west of Ventnor and up the west coast using Chain Home Low and Coastal Defence sets. More Ground Control Intercept stations had been installed and a number of night fighter aeroplanes had been fitted with the Air Intercept Mark IV sets. this was the position towards the end of October 1940. Electronic devices were beginning to play a very important part in the war and at this stage we were of the opinion that we had the lead over the Germans since they had not attempted to jam the Chain Home defence system. We assumed they did not know how to, other than by attacking the transmitter towers. All the Chain Home stations were now repaired and working.

Chapter 11

Royal Navy Become Involved

ABOUT THE END OF October 1940 Dr Dodds said that he had received an enquiry from the Royal Navy at Portsmouth who wanted us to produce some radio location equipment for naval ships. The basic design had already been done and we were to produce drawings and make the equipment as we had done with the radio location sets now working. Dr Dodds suggested that Mr Chadwick and I should go along to Portsmouth and find out what was wanted. We were to avoid getting involved in any naval numbering system similar to the Royal Aeronautical Establishment numbering system described earlier. We were to obtain the necessary information to produce the drawings and get the equipment made. Since there were air raids most nights on London and the south coast towns, it was decided that we should take tin hats and gas masks on what would be a four-day trip.

We set off for London where we had to change trains. I had never been to London but Mr Chadwick had lived there and said that if we had some time to spare he would show me around. We were not sure how the trains were running because of the nightly air raids, but when the first train was on time, Mr Chadwick decided that we could spend a few hours in London before we caught the train to Havant where we were to stay. We went along to Whitehall and to Downing Street where a small crowd of people were gathered. Somebody said Mr Churchill was due to go to the Houses of Parliament so we joined the crowd. He came out of Number 10 and everyone – even Mr Chadwick who was a 'dyed in the wool' Socialist – gave him a cheer to which he responded with a wave. We then walked to Waterloo station passing the Houses of Parliament and crossing Westminster Bridge en route before taking the train to Havant where hotel accommodation had been arranged for us.

The next morning a Royal Navy car with a WREN (Women's Royal Naval Service) driver came to the hotel for us and took us to the Signal School Radio Communications buildings which were not very far from Havant. The buildings were mainly large wooden huts which looked fairly new; and when we were taken into one of them it proved very similar to our Valve Lab. A variety of radio equipment was either under construction or being tested. We were shown the radio locators for which we were to make drawings prior to producing them; there were two different types: one was to work on a 1.25 metre wavelength (240 megacycles) with an output of about 10 kilowatts; the other was to work on a 50 centimetre wavelength (600 megacycles) with an output of 5 kilowatts. Both sets had to have accurate range-finders.

At this time in World War II the Navy's requirements for radio location sets differed from those of the Royal Aeronautical Establishment whose main interest was to locate enemy aeroplanes and shoot them down. While the Navy was also interested in locating aeroplanes chiefly for defence purposes, their main interest was to locate surface ships at night and in fog and better still to locate enemy submarines on the surface at night and in fog and sink them before the enemy knew what had hit them. The number of ships sunk in the Atlantic and in the approaches to the British Isles by the U-boats had been increasing – the Germans had increased their production of U-boats and top priority for radio location sets had moved to the Navy.

From September 1939 up to October 1940 there had

been extensive development of valves and other devices which would oscillate and give a good output on wavelengths below one metre (300 megacycles). These devices were to have a big impact on World War II. (See figure 14.)

We were to make the transmitters for both the 1.25 and the 50 centimetre wavelength sets and also the range-finding mechanism. The receivers were to be made by some other manufacturer. Both transmitters were similar to the Coastal Defence sets we had made – there were three sheet steel cubicles each about 3 ft high, 2 ft wide and 2 ft deep. There was one for the power supplies to the transmitters and receivers, one for the modulator and oscillator valves and one for the aerial coupling unit. The cubicles could be bolted together in any order whether vertically or horizontally.

Although the range finders were mounted on the receiver control panel (which was to be made elsewhere), we had been asked to make them even though we had not made anything of this type before. The range finder was a train of gear wheels which were turned by a small handwheel, the final drive being attached to a large potentiometer which could turn through 300 degrees. On the same shaft as the potentiometer was a pointer which also turned through 300 degrees. A fixed scale was arranged under the pointer: at the left-hand end it was marked 0 and at the right-hand end it was marked 20,000 yards. Between these points the scale was calibrated in divisions of 1,000 yards. There were four gear wheels and a pointer was attached to each shaft with a fixed scale under each pointer; the scales for the first three gear wheels were marked 0 to 9. On the scale over the gear wheel which was turned by the hand-wheel was the figure 10, over the next wheel was the figure 100 and over the next, 1,000. The fourth gear wheel was calibrated from 0 to 20,000 yards (as explained).

The whole unit measured about 12 in. x 8 in. and was mounted on the receiver control panel. Above the range-finder was a 12 in. (30 cm) diameter cathode-ray tube, the

FIGURE 14

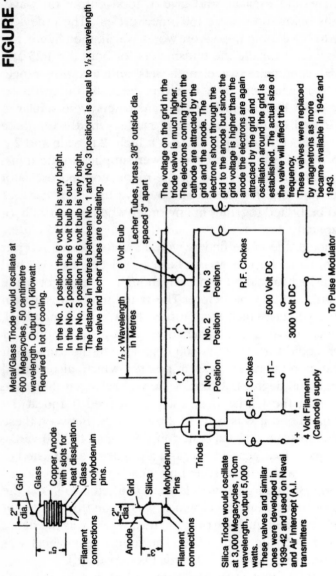

Metal/Glass Triode would oscillate at 600 Megacycles, 50 centimetre wavelength. Output 10 Kilowatt. Required a lot of cooling.

In the No. 1 position the 6 volt bulb is very bright.
In the No. 2 position the 6 volt bulb is out.
In the No. 3 position the 6 volt bulb is very bright.
The distance in metres between No. 1 and No. 3 positions is equal to ½ x wavelength the valve and lecher tubes are oscilating.

Lecher Tubes, brass 3/8" outside dia. spaced 3" apart

6 Volt Bulb

½ x Wavelength in Metres

No. 1 Position   No. 2 Position   No. 3 Position

R.F. Chokes

R.F. Chokes

HT−

5000 Volt DC

3000 Volt DC

To Pulse Modulator

4 Volt Filament (Cathode) supply

Triode

The voltage on the grid in the triode valve is much higher. The electrons coming from the cathode are attracted by the grid and the anode. The electrons shoot through the grid to the anode but since the grid voltage is higher than the anode the electrons are again attracted by the grid and oscillation around the grid is established. The actual size of the valve will affect the frequency.

These valves were replaced by magnetrons as more became available in 1942 and 1943.

Grid
Glass
Copper Anode with slots for heat dissipation.
Glass molybdenum pins.
2" dia.
5"
Filament connections

Grid
Silica
Molybdenum Pins
Anode
2" dia.
3"
Filament connections

Silica Triode would oscillate at 3,000 Megacycles, 10cm wavelength, output 5,000 watts.
These valves and similar ones were developed in 1939-42 and used on Naval and Air Intercept (A.I. transmitters

TRIODE VALVES USED ON CENTIMETER WAVELENGTH TRANSMITTERS

largest cathode-ray tube I had ever seen. It looked huge
compared with the 6 in. (15 cm) and 9 in. (23 cm) tubes we
had used so far. On the receiver control panel there was
another handwheel with a scale marked 0)' at the top and
with the left-hand side marked 'Port' and divided into 180
degrees and the right side marked 'Starboard' (and also
divided into 180 degrees). This handwheel turned a selsin
unit: as the handwheel was turned the selsin motor attached
to the handwheel shaft drove a selsin slave motor through
the same angular movement as the handwheel selsin unit
turned. The selsin slave motor was attached to the
transmitter and receiver aerials so that the aerials could be
turned by almost 360 degrees. By turning the handwheel
with the selsin unit attached, one way and another the
whole area for 20,000 yards around the ship could be
searched for surface vehicles, submarines on the surface or
low-flying aeroplanes in daylight, at night and in fog.

The monitored transmitted pulse was displayed at the
left-hand side of the cathode-ray tube as a spot of light;
another spot of light was also displayed just above it. This
was from the range finder. When the two spots coincided,
the rangefinder was reading zero on all scales. As the
aerials scanned the sea around the ship, if there was
another ship or aeroplane within the 20,000 yard range, a
spot of light would appear at the right-hand side of the
cathode-ray tube. When this happened the handwheel
driving the selsin unit had to be adjusted very slowly to
keep the spot of light as bright as possible. Then the handwheel controlling the rangefinder had to be turned as fast
as possible to bring the range finder spot of light in line
with the spot of light at the right-hand side of the cathode-
ray tube. By adjusting the two handwheels, the target range
in yards and the target's position relative to the ship could
be continuously monitored.

The second range finder we had to make was identical
in many ways to the one I have just explained, the main
difference being that it was to have a range of 0 to 40,000

yards – the fourth gear wheel and the scale would be different. This arrangement of radio location equipment which the radio section of the signal school had produced was the best we had encountered up to that time. They had the edge on any navy in the world and the development was most secret.

Mr Chadwick and I returned to the hotel in Havant and later that evening decided to have a look at some of the standard specifications we had been given. They were of a general nature and not secret in any way. I had noticed that the steel work, of which the chassis, cubicles, framework and metal supports were made, was much heavier than we had used on Chain Home Low and Coastal Defence transmitters. I pointed this out to Mr Chadwick who had also noticed the thickness of some of the steel work they had used. As we looked through the standard specifications we found the reason. One specification was headed 'Electrical and Radio Equipment Drop Test'. It detailed metal-clad switchgear, fusegear, motor controllers, junction boxes, lighting fittings, radio chassis and frameworks, and went on to say that all containers enclosing any of the above equipment were subject to a drop test detailed as follows. Each unit had to be lifted to a height of 30 in. and then dropped on concrete so that one corner of the unit hit the concrete first. Throughout this test it was essential that the container was not distorted, that the doors, panels and gaskets were not bent or broken and that the contents were not damaged. Radio valves and indicator lamps were not included in the test but ammeters and voltmeters were. The other specifications concerned fixings and the locking of screws and nuts and bolts with 'lok-tight' glue. All soldered joints were to be joined mechanically first and then soldered. There were painting and electro-plating specifications and all timber had to be protected against every known grub which had a liking for it. I suppose the Navy had built up these specifications after years of experience in all parts of the world.

During the night I heard the air raid sirens in the

distance. There were a few bangs but nobody in the hotel seemed bothered so I supposed this was a regular occurrence. Next day we went back to the Signal School with a list of queries for the two men we had seen the day before. We also wanted more details about the drop test specification – was the equipment we were going to make to be subjected to it? The answer was affirmative. We could not understand the reason for the test until it was explained: when the larger guns were fired or there was a near miss by enemy shells the ship would vibrate and anything not properly secured would break. The Navy must have learned this lesson from bitter experience.

One of the radio location sets was designated for training operators and we were given a demonstration of how it worked, using a simulated target. The range could be ascertained down to one yard by reading the pointers shown on the scales. If the pointer was between two numbers the lower number of the two was read, in the same way that a gas or electricity meter is read. We pointed out that the drawings we would make for all the parts we were to produce would have one drawing and one drawing number for each item. If they had their own numbering system, they could add their numbers later by themselves. This was agreed. I think they were happy to get somebody interested in producing in quantity the equipment they had designed.

Since the radio location sets were top secret, drawings, diagrams and specifications would be sent through the official channels to Dr Dodds. We could not take any paperwork with us. We returned to the hotel in Havant and the next day went back home. When Mr Chadwick and I arrived in the Valve Lab the next morning, Dr Dodds and Mr Ludlow were waiting for us to find out how we had got on and also to find out details of what we had to make. They were a bit disappointed when they realised we had no drawings or specifications.

A few days later a parcel was delivered to the Valve Lab door and inside were the drawings, diagrams and specifications

FIGURE 15

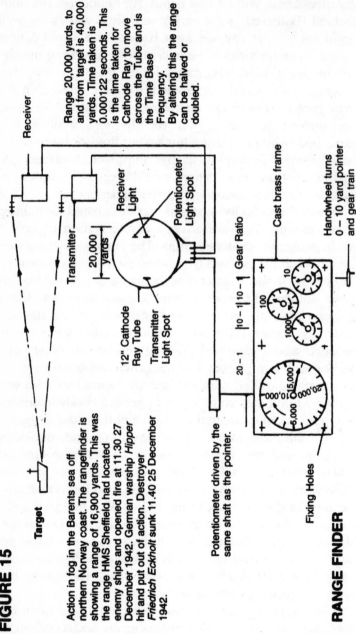

Target

Action in fog in the Barents sea off northern Norway coast. The rangefinder is showing a range of 16,900 yards. This was the range HMS Sheffield had located enemy ships and opened fire at 11.30 27 December 1942. German warship *Hipper* hit and put out of action. Destroyer *Friedrich Eckholt* sunk 11.40 25 December 1942.

Receiver

Transmitter

Range 20,000 yards, to and from target is 40,000 yards. Time taken is 0.000122 seconds. This is the time taken for Cathode Ray to move across the Tube and is the Time Base Frequency.
By altering this the range can be halved or doubled.

Receiver Light

Potentiometer Light Spot

20,000 yards

12" Cathode Ray Tube

Transmitter Light Spot

Potentiometer driven by the same shaft as the pointer.

Gear Ratio

20 – 1 |10 – 1|10 – 1

Cast brass frame

10

100

1000

15,000

10,000   20,000

6,000

Handwheel turns 0 – 10 yard pointer and gear train

Fixing Holes

RANGE FINDER

for the two radio location transmitters and the two different rangefinders we were to make. The order was for ten 1.25 metre wavelength transmitters Type No. 281, twenty 50 centimetre wavelength transmitters type No. 282, fifteen 0-40,000 yard range finders and twenty-five 0-20,000 yard rangefinders. Dr Dodds and Mr Ludlow were quite surprised since this was the first order we had been given for shipborne radio location equipment. Mr Ludlow said we must have impressed the admirals. Mr Chadwick explained what we had seen and related that to the drawings and specifications we had received. After some discussion it was decided that cast brass frames would be best for the rangefinders. The cubicles would be ³/₁₆ in. thick sheet steel of welded construction; the steel chassis carrying components inside the cubicles would be ¹/₈ in. thick; the chassis in the high frequency chambers would be ¹/₈ in. thick aluminium; all connections between cubicles would be by plugs and sockets counted on the front panels; and ammeters, voltmeters and indicators were to be industrial types with wired glass. All cubicles were to be drip proof and hose proof. The finish was battleship grey paint. We were of the opinion that this arrangement would stand up to the drop test specification. I was to make the drawings for the rangefinders, Mr Chadwick was to make the drawings for both transmitters helped by the draughtsmen in the new drawing office, Mr Ludlow would be responsible for overall design and if there were any hold-ups in obtaining components, Dr Dodds had to be informed. (See figure 15).

I made the drawings for the cast brass frames. They were about 8 in. wide, 12 in. long and ¹/₂ in. thick and had raised bosses where the spindles for the gear wheels and pointers went through the casting. These bosses were machined. There were also six machined bosses for the fixing holes. It had been decided that we could justify drilling jigs for both castings so all dimensions were on X and Y co-ordinates from a hole near the centre of the castings. I gave the pattern-makers and the tool room advance copies of the drawings so that they could get started. The gear wheels were a standard type and just had to be ordered. We fitted 'oilite' bearing bushes for the spindles; the labels and scales were made of 'traffolyte', and the potentiometers were as detailed in the specification sheets we had received. But in November 1940 we still had to produce and assemble all the components and get them to work correctly.

The Luftwaffe were now bombing towns and cities most nights and as the nights grew longer, they were penetrating farther to the north and west of the country. About the middle of November 1940 Coventry was badly bombed and it was fairly obvious that the German air force had used their radio beams to aim the bombs because of the accuracy of the attack. By examining a bomber which had been shot down and using radio test equipment developed for analysing wave forms, we had found that they were using wavelengths between 2 and 5 metres. The German transmitters were sited in the Brest and Cherbourg areas of France. Most of our transmitters were pulsed and working on 1.5 metres. Radio transmitters were needed which worked on continuous wave to cover the 2 to 5 metre band of wavelengths to jam the German transmitters. Some of the 1.5 metre wavelength transmitters we were now producing could be modified and used as 'jammers'. This was the start of the development of 'electronic counter measures' which became so important to both ourselves and the enemy as World War II progressed. Towards the end of the war 'electronic counter-counter measures' were produced.

I got married on Saturday 16 November 1940. We did not have a honeymoon as I was too busy and was back at work at 9.00 a.m. on the Monday morning. My wife and I rented a house in Swinton, Manchester at a cost of 17s. 6d. (87p) per week and it had a garden. Everybody was asked to 'dig for victory' by growing vegetables on their garden plots, but since it was November all I could do was to turn over the soil. I thought that some of Mr Thornton's hen droppings might do the soil some good, but he could not supply any now that the petrol ration for private motorists had been stopped.

By December 1940 a number of the Air Intercept sets had been fitted to aeroplanes. These performed well and were less affected by the array of aerials sticking out in front. More Ground Control Intercept (GCI) sets had been installed and some new searchlights had been fitted with radio locators so that they could follow enemy aeroplanes. All this was having some effect on the number of enemy night bombers shot down, but not yet enough to make it too expensive for the Luftwaffe to continue.

About the middle of December 1940 the Ministry of Information issued a small book which cost 2s. (10p) about how the war was progressing and giving some details about the air battles that had taken place between June and October 1940 over the south and east coasts and over London. It was mainly a morale booster and was very well written. It gave estimated enemy aircraft losses of about 2,000 and detailed our own losses at 700 aeroplanes. It also gave some details of the tonnage of shipping lost and estimated the number of U-boats and enemy surface ships sunk or damaged. There had been a bumper harvest and the 'dig for victory' campaign had proved very successful. One of the experts from the Ministry of Agriculture had worked out that more vegetables had been grown per acre in the gardens of houses than would have been the case if the land was just farmland. This was before the age of computers. Towards the end of the book there was a small paragraph which said, 'As a result of using some secret equipment which had been developed, we were able to locate aeroplanes in cloud and at night and this had helped to defeat the Luftwaffe'. The secret was out. Up till then it had been the best kept secret of World War II.

In their newspapers some of the foreign correspondents had been reporting that the British had a secret weapon and some of our daily newspapers had reported what was appearing in the foreign press. I showed the paragraph to my wife and told her that this was what I had been working on since 1936. She had no idea until I told her.

At about 8.00 p.m. on 22 December 1940 the air raid sirens began to wail so I went outside to see if there was anything happening. There were a number of parachute chandeliers coming from the south-west and going towards Manchester; it was like daylight. Within a short time the Germans were dropping incendiary and high explosive bombs on the centre of the city-which was soon on fire. It was still on fire when the bombers returned on the nights of 23, 24 and 25 December.

The Manchester Water Works had a system of water hydrants in every part of the city for use in case of fire. Each hydrant had a ball valve which was held in position by the water pressure. This arrangement worked all right until the German bombs broke some of the water mains which reduced the water pressure so that all the ball valves dropped from their seats and released any water left in the pipes. Consequently there was no water to put out the fires. On the night of 24 December 1940 Trafford Park and Manchester Dock were bombed and Met-Vick also received its share of incendiaries and high-explosive bombs. The Valve Lab and the Research Department escaped with some broken windows but the main factory roof caught fire. It was made of glass and timber covered with roofing felt. As in the centre of Manchester, the water pressure fell and released the ball valves and the water just trickled away. The water tower which was taken down in January 1940 would have been very useful at the end of that year.

On the same night the air raid sirens sounded and we could hear some activity in the distance. All of a sudden there was a noise like an express train and a series of very loud bangs. The whole house seemed to rock. Both my wife and I dived under the table in a flash. All was very quiet for about half an hour when we heard somebody shouting to us to go to the air raid shelter which was in the basement of a church just near to where we lived. It was then about 1.00 a.m. on Christmas Day 1940. We were told that there was a suspected unexploded bomb somewhere in the area and until it was light we should remain in the shelter. At about 1.00 p.m. we were allowed to go back home – the bomb could not be found. We had no water, gas or electricity and no Christmas dinner.

On 27 December 1940 I was back at work in the Valve Lab. The damage to the main factory roof was very bad and as a temporary measure it was covered with tarpaulins. Luckily the effect on the various radio transmitters and parts we were now making was not very great and very soon we were back to full production. As a result of the fire damage, which had been mainly caused by incendiary bombs, it was decided to enlarge the auxiliary fire service in the factory and have pumps and water supplies independent of the hydrants dotted around the various buildings. A request for men to join the works auxiliary fire service was circulated to all departments. I volunteered and was given a uniform, a fireman's tin hat and a gas mask. We were arranged in units of seven men (one leader and six firemen) and were allocated a Coventry Climax fire pump which we had to man-handle and use for fire fighting. We had an eight-day rota and had to be on call for 24 hours every eight days. We needed a lot of training. I was still a member of the Local Defence Volunteers, but I think they had lost track of me since I had changed my address when I got married. Also, the threat of invasion seemed to have passed.

Following the appearance of those few lines regarding secret equipment in the Ministry of Information booklet, a number of people from the various Ministries and Services came to see who we were and what we were doing. Dr Dodds and Mr Ludlow showed the visitors around the factory where parts were being made and assembled and sometimes Mr Chadwick and I got involved when drawings and specifications were required to explain the operation of the various parts. Some of the visitors ended up in the

Valve Lab to see a demonstration of Dr Dodds' 'All Singing, All Talking Arc' and to hear Mr Ludlow giving a rendering on his tone generator, closing with the National Anthem. We had a visit from some American generals. When Dr Dodds' 'singing arc' was demonstrated to them, one of them wanted to buy one but Dr Dodds told him he would have to wait until the war ended.

Chapter 12

New Developments

IT WAS NOW January 1941. During 1940 there had been a number of very important developments made in the design of radio location equipment. A metal-glass valve was now being produced which would oscillate at 3,000 megacycles, 10 centimetre wavelength with an output of about 1,000 watts. It was a triode about 5 in. high and 2 in. diameter. The anode in the centre was made of copper, machined with slots to radiate the heat. Glass caps with the filament and control grid connections were fitted one at each end of the copper anode and sealed by a new fusion technique. A unit had been developed which enabled a common aerial to be used for both transmitting and receiving. It passed weak signals to the receiver but blocked the transmitter power pulses from it.

The Chain Home Low Transmitter and receiver aerials had been arranged to turn through 360 degrees either way. They were power-driven and one revolution took about 60 seconds. By turning the aerials backwards and forwards an area of about 60 miles radius from the aerials could be swept. It was noticed that all aeroplanes in that area were located and a spot of light appeared along the left-to-right trace on the cathode-ray tube for each one, showing the distance each aeroplane was from the aerials. The position, north, south, east or west of the aerials was also known by the position of the aerials when the spots of light were very

bright. The spots of light would flash for friendly aeroplanes so long as the Friend-Foe Indicators were switched
on. They would not flash if the aeroplanes belonged to the
enemy.

From these observations the Plan Position Indicator (PPI)
was developed. Now that just one aerial was required for
transmitting and receiving it was arranged so that it could
turn continuously at about four revolutions per minute.
Attached to the aerial drive shaft was a selsin unit which
also turned at 4 rpm. This drove a slave selsin unit which
followed the one attached to the aerial. The slave selsin
unit was arranged to turn the deflection coils around the
neck of the cathode-ray tube, and the trace made by the
electron beam in the cathode-ray tube was arranged so that
it appeared to start at the centre of the tube and move
outwards. With this arrangement, as the aerial turned and
swept a 60 mile radius the beam of this cathode-ray tube
swept the area of the tube face – the distance from the
centre of the tube to the outside edge also represented 60
miles. When any aeroplanes came within the sixty mile
radius they were located. Their position and distance away
were continuously shown on the cathode-ray tube. The
height of the aeroplanes was shown on another cathode-
ray tube. The first Plan Position Indicators were designed to
use the 1.5 metre (200 megacycle) transmitters which we
were now producing in large quantities.

A device called a magnetron had been developed at
Birmingham University which would oscillate at 3,000 megacycles, 10 cm wavelength, and could produce up to 10 kilowatts of power. The principle of the magnetron had been
known from about 1930 but it had not been developed
because it generated such short wavelengths. There appeared
to be no use for such a device because the ultra-short
wavelength technology in the 1930s was not advanced
enough to find a use for them. Things had changed in 1940
mainly because of the pressure of World War II for
advanced electronic devices. I still have a book – *Modern*

Radio Communications Volume II (1935) – which refers to the magnetron as an ultra-short wave oscillator in which the movement of the electrons is controlled magnetically instead of electrostatically. The frequency which the magnetron generated was determined by its physical size and the strength of the magnetic field which accelerated the electrons in a circular path around the anode cavities and generated short pulses of power. The magnetron was just what was required for use in centimetre wavelength radio location transmitters. (see Figure 16.)

In January 1941 the magnetron was still being used for trials and was not yet in production. The developments during 1940 which I have referred to were to have a big effect on the electronic warfare which was now beginning between our scientists and German scientists for mastery in the air. It had been decided that the Navy was to have top priority for any equipment we were making because shipping losses were increasing as a result of greater activity by German submarines, mainly in the Atlantic and the sea lane approaches to the British Isles. As a result of this decision some of the Air Intercept sets on 1.5 metre wavelength which were being fitted to night-fighter aeroplanes were diverted to Coastal Command for fitting in their aeroplanes and became known as 'Air to Surface Vessel Mark II sets'. These sets would be helpful for locating convoys at night and in fog and also for locating and attacking German submarines on the surface, also at night and in fog. On the larger Coastal Command aeroplanes 'side looking' sets were fitted and by using these an area of about seven miles radius from the aeroplane could be covered.

We received an order for another 30 airborne transmitters and power supplies. The German air force was still sending bombers over most nights and with the long hours of darkness they were able to range far and wide. The jamming equipment we were now using to jam and distort the radio beams they were using to locate targets was

FIGURE 16

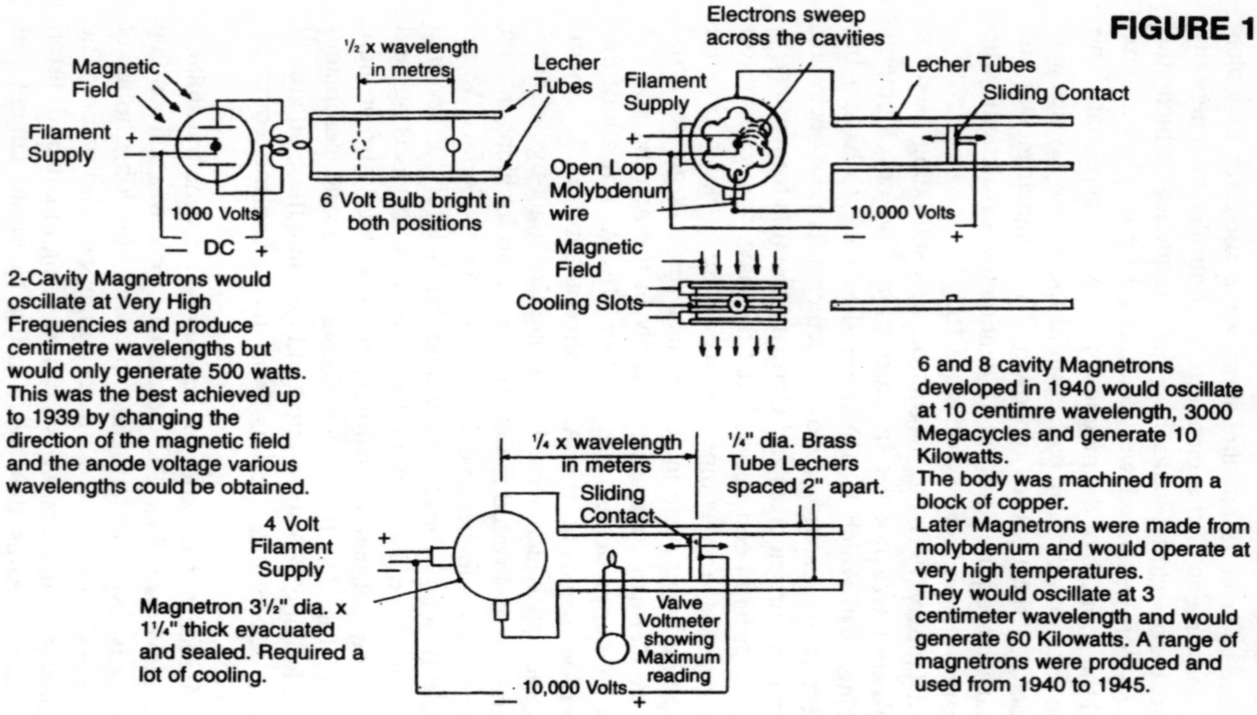
Magnetic Field
Filament Supply
+
—
½ x wavelength in metres
Lecher Tubes
1000 Volts
— DC +
6 Volt Bulb bright in both positions
2-Cavity Magnetrons would oscillate at Very High Frequencies and produce centimetre wavelengths but would only generate 500 watts. This was the best achieved up to 1939 by changing the direction of the magnetic field and the anode voltage various wavelengths could be obtained.
Electrons sweep across the cavities
Filament Supply
Lecher Tubes
Sliding Contact
Open Loop Molybdenum wire
10,000 Volts
Magnetic Field
Cooling Slots
6 and 8 cavity Magnetrons developed in 1940 would oscillate at 10 centimre wavelength, 3000 Megacycles and generate 10 Kilowatts.
The body was machined from a block of copper.
Later Magnetrons were made from molybdenum and would operate at very high temperatures.
They would oscillate at 3 centimeter wavelength and would generate 60 Kilowatts. A range of magnetrons were produced and used from 1940 to 1945.
¼ x wavelength in meters
¼" dia. Brass Tube Lechers spaced 2" apart.
Sliding Contact
4 Volt Filament Supply
Magnetron 3½" dia. x 1¼" thick evacuated and sealed. Required a lot of cooling.
Valve Voltmeter showing Maximum reading
10,000 Volts

MAGNETRONS USED ON CENTIMETRE WAVELENGTH TRANSMITTERS

having some effect. By listening to the two-way radio conversations between the German pilots and the ground controllers in France it was concluded that they were blaming each other and also their radio beam equipment for not being able to find the target. It had been hoped that this would happen so that they would lose faith in their beams. At the same time the Ground Control Intercept (GCI) stations and the night-fighters with their Air Intercept Mark IV sets had been getting more used to the radio location equipment, and every time there was a raid a number of German bombers were being located and shot down or damaged. Anti-aircraft guns were sited near or around most possible targets but were not very accurate. They would put up a barrage of shells in the general direction of any aeroplane which came into their space in the sky and they were a good morale-booster for the people being bombed.

There had been an earlier attempt at anti-aircraft gun control using radio location on a 5 metre wavelength but it was not very successful. This was called Gun Layer (GL) Mark I. What was required was a radio location set on a much shorter wavelength coupled to an electronic predictor which could predict the location of an aeroplane by the time the shell had travelled to the target. This was assuming that the aeroplane kept flying at the same height and speed and in the same direction. Such a device had now been developed and became Gun Layer (GL) Mark II. We received an order for 20 transmitters to operate on a 50 cm wavelength (600 megacycles) with an output of 10 kilowatts. The transmitters were to be mobile-housed in trailers similar to the Coastal Defence sets we had made. All the transmitters we were now making for the shorter wavelength sets were in three cubicles or boxes. There was a power unit, a modulator-oscillator unit and an aerial coupling unit. The units were connected to each other by plugs and sockets usually mounted on the front panels. The transmitter aerials were to be mounted on top of the trailer and the turning mechanism would be inside the trailer. Gun Layer Mark II was designed as an anti-aircraft gun controller for ground to air attack. From this a Gun Layer was developed for ground to ground attack (artillery).

So far, the process of finding out the direction and height of aeroplanes by the Chain Home stations had relied on at least two stations locating the aeroplanes. By using the information from the stations an estimate of the direction and height could be calculated. By taking a series of readings the track and height of the aeroplanes could be plotted. A new device called a 'goniometer' had been developed which gave the direction and height very accurately for each station. The direction and height could be displayed on cathode-ray tubes on a continuous basis without any calculation. The Chain Home Low sets were modified in a similar manner.

As a result of these modifications and some other alterations, Mr Ludlow, two technical assistants and I were to pay a further visit to the Chain Home site at Danby Beacon. I was to take some blueprints and diagrams and mark them with the alterations which had been carried out. Towards the end of March 1941 we set off in Mr Ludlow's car for north Yorkshire. All road signposts had been removed in 1940 when there was a threat of invasion so we had to work out the way by map reading. I was in the back seat and did not take any part in the direction finding. En route Mr Ludlow told us a story about RAE Farnborough Design Section. They had decided to take part in the design of airborne radio location sets. In all the airborne electronic equipment we had designed so far we had kept the valves cool by using small centrifugal airblowers driven by a 24 volt DC motor taking about 50 watts supplied from the 24 volt aircraft battery. These airblowers added to the overall weight of the equipment being installed in the aeroplanes and the RAE design section decided to reduce the total weight by using the air flowing past the skin of the aeroplane to cool the valves. They put some metal tubes in place of the centrifugal airblowers, with a rubber tube connected to the metal tubes. The rubber tube was taken outside the aeroplane so that the forward movement of the aeroplane forced air down the tubes and across the valves. The design also ensured that rainwater did not get into the tubes and the arrangement saved the weight of the centrifugal airblowers. The new system for cooling the valves worked very well until the aeroplane went through a cloud full of moist air and the water-laden air found its way into the valve chambers and other electrical components. We carried on using motor-driven centrifugal air blowers.

Once again we stayed at the Sandsend Hotel. There had been a great change since our last visit in 1940. The beach was now covered with barbed wire and posts were stuck in the sands and there were also a number of pillbox defences and gun emplacements. The next morning we went along to the Chain Home station to discover that the security had been improved but that otherwise it was just as it was in 1940. One of the transmitters had been shut down and Mr Ludlow and the two technicians had to carry out some modifications and adjustments to the wiring and connections between the control desk and the two transmitters. My job was to keep a record of what was being done. By about 6.00 p.m. we were back at the hotel. We repeated this exercise the next day when we carried out the modifications to the other transmitter, checking beforehand that the first transmitter was on air and working satisfactorily. Next day we called at the site to make sure both transmitters were in working condition and after about two hours we were on our way home. The modifications were to reduce the effect of any jamming of the transmitted and received radio pulses by the enemy. I modified the drawings in line with what had been done at Danby Beacon and issued the new blueprints and instructions so that the other Chain Home transmitters could be modified in a similar manner.

Radio transmissions on about 2-metre wavelength (150 megacycles) had been detected coming from a number of

places along the French coast. The sites had been located
and aerial photographs had shown aerial arrays similar to
our own Chain Home Low transmitting and receiving
aerials. It was very obvious that the Germans had developed some sort of radio location equipment of their own
and were able to detect our shipping and aeroplanes along
the south coast. This was a bit of a blow to our pride. Up
to now we had assumed that we alone had developed an
operational radio location system. The German transmissions on the 2-metre wavelength were however very weak
and the detection range would have been only about 30
miles. Now, there would have to be a constant watch on all
German radio transmissions, mainly to keep the lead we
had in this field and also to develop jamming equipment to
be used when it became necessary to do so.

The German air force had been carrying out bombing
raids on most nights, but as a result of the improved
Ground Control Intercept sets and the Air Intercept Mark IV
airborne sets the enemy had been losing a large number of
aeroplanes, making the raids very expensive in pilots and
navigators. In May 1941 the Ministry of Information issued
another small book, cost 2s., (10p) which gave more details
about how the war was progressing. The tonnage of
shipping lost resulting from U-boat action was increasing
but the number of U-boats sunk or damaged was also
increasing. Nightly bombing raids by the German air force
had continued, but about 10 per cent of all German night
bombers were now being shot down or damaged. The
reason given for this was that the RAF night-fighter pilots
were being fed on carrots which helped them to see in the
dark. There was no mention that the secret equipment
referred to in the previous booklet had now been fitted in
night-fighter aeroplanes and naval ships. For many years
after the war was over quite a lot of people were still of the
opinion that eating carrots helped you to see in the dark. It
was very good propaganda – there was probably a surplus
of carrots in May 1941.

In May 1941 the Germans invaded Russia. The Luftwaffe now had other targets to bomb and as a result of the invasion of Russia and the shorter hours of darkness, the nightly bombing raids on British towns and cities was reduced. In the same month the German battleship *Bismarck* broke out into the Atlantic Ocean to try to sink shipping coming from America and other parts of the world. Some of the naval fighting ships had been fitted with radio location sets which helped to locate the *Bismarck* before it was sunk on 27 May 1941.

A new Friend-Foe Indicator had been developed and we were to make the drawings and specifications and also manufacture them. They were very similar to the earlier ones we had made, but were to operate on 150 to 200 megacycles (1.5 to 2 metre wavelength) for receiving and replying and were to be called the Friend-Foe Indicator Mark III. They became standard during World War II and were used by all Allied air forces and also by the Royal Navy. We must have produced several thousand FFI Mark III sets. This was now becoming the pattern. The radio location sets we had designed and produced were all being constantly updated, revised and replaced with modified and new sets. Each revised set was given a new number or a new mark number and these new numbers had to be included on the drawings and specifications. Sometimes similar parts were used for different mark numbers.

The Chain Home, Chain Home Low and Coastal Defence sets had all been fitted with cathode-ray tubes for direction and height-finding. The Ground Control Intercept sets had been modified to Plan Position Indicator (PPI) sets with a range of about 60 miles all round the common rotating transmitting and receiving aerials. More night-fighter aeroplanes had been fitted with the airborne Air Intercept Mark IV sets and a number of Royal Navy ships had been fitted with the 1.25 metre wavelength Type No. 281 sets or the 50 cm wavelength Type No. 282 sets. Some of the larger ships were fitted with both types. RAF Bomber Command had been making nightly bombing raids on targets in Germany, but the results were not very good as the bombs were not landing on the targets. A navigation system similar to the German Beams or an improved system was required that would be very difficult to jam. Quite a large number of electronic components were now coming from the United States and Canada and were being used in ever increasing quantities as the demand for all kinds of electronic equipment that we were designing and manufacturing increased. This was the position in the business of electronic warfare in July 1941.

We had received an initial order for 20 improved airborne Air Intercept transmitters and Dr Dodds brought some sketches with outlines of the valves to be used into our office for Mr Chadwick and me to study. They were completely different to anything we had made so far. There was a parabolic-shaped dish of about 30 in. diameter; the valves were triodes with a glass envelope about 5 in. high and 2 in. diameter with short stub connections about $1/8$ in. diameter coming straight out of the glass envelope. The tuning was to be carried out by two Lecher tubes each about $3/8$ in. outside diameter and 10 in. long, spaced 3 in. apart and supported on some new insulation material which had been recently developed and given a War Department (WD) reference number. With this arrangement the transmitter was to operate on a 10 cm wavelength and a frequency of 3,000 megacycles with a pulsed power output of about 5,000 watts. We were to supply the power units, transmitters and parabolic-shaped dishes and were to provide power supplies for the receivers and cathode-ray tubes. The receivers, cathode-ray tubes, aerials and the common transmit-receive units would be supplied by others.

These radio location sets were the shortest wavelength (highest frequency) we had made so far and were to be Air Intercept Mark VIII. The maximum range would be about 25,000 ft, the minimum range 300 ft and they would be very accurate in the range 10,000 ft to 300 ft. Up to 1940 there was a general belief that long distance communication on short waves ceased at about 5 metre wavelength (60 megacycles) and that frequencies above 60 megacycles were of little use. Wavelengths of 1 metre (300 megacycles) and shorter wavelengths would travel in straight lines and would not bend around the surface of the earth. If there was any obstruction in the path of the transmitted waves they would be completely blocked. This meant that in order to transmit such waves over long distances the aerials would have to be very high. There was also a question of all kinds of interference from electrical machinery and motor car ignition systems which affect transmissions in the 1 metre to 10 metre band of frequencies.

On the first 405 line television sets the picture would be completely blocked out by a passing motor car. Now that valves had been developed which would oscillate at 3,000 megacycles, 10 cm wavelength, with a reasonable output the radio waves developed could be directed in a narrow beam using a parabolic reflector similar to a searchlight and, like a searchlight, the radio waves would rebound from any object in their path. These reflected radio waves could be detected, amplified and displayed on a cathode-ray tube in the same way that they were with the longer wavelength sets. By turning the parabolic reflector, a large area could be swept (the process was similar to the operation of a searchlight). The distance to the target shown on the cathode-ray tube was very accurate and the transmitter and receiver would be very difficult to jam.

We had an order for 20 of these new airborne transmitters but Dr Dodds said he would like four as soon as possible so that he could carry out some tests. Mr Chadwick decided it would be best if I did the drawing for the parabolic-shaped dish for which the curvature had to be developed from a formula given to us by Dr Dodds. I decided to make the drawing full-size and by plotting the co-ordinates on the X and Y axis the shape of the dish was developed. Dr

Dodds had decided that the outside diameter would be about 30 in. and at this diameter the dish was about 5 in. deep. Dr Dodds had also said that the dish must be smooth and polished on the inside surface. I had no idea how these dishes would be made so I took a blueprint along to the Sheet Metal Shop and explained to the foreman what was wanted. I pointed out that the drawing was full size and if he wished, he could cut the drawing to ensure that the correct curvature was obtained. We could not use sheet steel as the metal had to be non-ferrous – with copper, brass or aluminium. When I asked him how he proposed to make them, he said it would be done by panel beating. He also said he had some $1/16$ in. (1.5 mm) thick copper sheet that would be suitable. I left the drawing with him and told him an order for 20 dishes would follow, but I would like four dishes as soon as possible. I also pointed out that the inner surfaces of the dishes must be polished and free from bumps and blemishes. The idea of panel-beating bothered me.

When I returned to the Valve Lab and explained to Mr Chadwick how the dishes were going to be made, he was not very impressed and told me to ring the foreman and tell him to make only four dishes and no more until we had checked the finish. Mr Chadwick explained to me that panel-beating was used on motor cars that had been in an accident to put them back into shape, but all the bumps and scratches were filled in and painted over. This was not acceptable for the dishes.

The next day the foreman of the Sheet Metal Shop rang me to say that the four dishes were ready for collection. I told Mr Chadwick and we both went along to have a look at the finished product. We need not have worried – the sheet-metal workers were masters of their craft. The copper dishes were brightly polished and had a slightly mottled finish. Mr Chadwick said they were too good to go on aeroplanes and would be better as plant-pot holders in his house. He later had a go at this panel-beating himself and became very good at it.

The other aspect that was different from any other radio location sets was the method of tuning, which was done by using two Lecher tubes which were $3/8$ in. outside diameter brass tubes spaced 3 in apart and mounted on blocks of the new WD insulation material which was very similar to perspex. It could easily be machined to shape, drilled and tapped, and was most suitable for insulating components carrying ultra-high frequency currents. A brass bridging piece was arranged to slide along the Lecher tubes and was clamped in position where the maximum power was generated. This point on the Lecher tubes was found by replacing the brass bridging piece with a filament lamp, the connections of which could be slid along the two tubes. As the lamp was moved along the tubes, the filaments would become very bright, then go out and become very bright again before going out once more. These points were marked and measured accurately in centimetres which gave the wavelength.

On the drawings it was specified that the Lecher rods, brass bridging piece and connection clamps to the Lecher rods were all to be Rhodium (flash) electro-plated. This was mainly to reduce the possibility of corrosion; since high-frequency currents flow on the outside of conductors it would help in the passage of the oscillating currents around the circuit. One Lecher rod was connected to the control grid of the valve and the other was connected to the anode of the valve. The 5,000 volt DC power supply was fed into the system via four high frequency chokes. Two were in the positive lines to the valve grid and anode, connected to the Lecher rods at the voltage node or the point where the filament lamp was out. The other two high frequency chokes were connected to the valve filaments (cathode) connections and the 4 volt filament supply positive was connected to the 5,000 volt DC negative, thus forming two circuits through the valve. The voltage of the grid was kept constant at 5,000 volts while the anode voltage was supplied via the pulse modulator up to about 3,000 volts.

At this stage in the development of the technology it was generally thought that the electrons produced by the heated filament (cathode) would be attracted by the grid and the acceleration of the electrons would be so great that a large number of the electrons would pass through the grid mesh and continue towards the anode. Since the grid voltage was much higher than the anode voltage, the electrons would decelerate and try to return to the grid. Some electrons would pass through the grid and join up with electrons coming from the filament (cathode) and in this way the valve would oscillate at about 3,000 megacycles on a wavelength of between 9 and 10 cm depending on the physical dimensions of the valve. By switching the anode voltage via the pulse modulator a series of ultra-high frequency pulses were generated. The output coils in the high frequency coupling box were connected at the point where the filament lamp was very bright.

It was a general belief that an oscillation of 3,000 megacycles was the maximum possible that could be generated because the electrons had reached their ultimate speed which could not be increased. The power unit, the oscillator and modulator unit and the aerial coupling unit were arranged in three sheet-aluminium boxes each measuring about 24 in. cube. This design for the Air Intercept Mark VIII set became the basic design for all of the 3,000 megacycles sets made. Since top priority had been given to the Navy we designed and produced Type 271 and Type 272 sets for use in the smaller vessels.

Chapter 13

Panoramic Transmitters/Receivers

SINCE THE DISCOVERY of the German transmissions on a 2-metre wavelength coming from the French coasts, we had been sending aeroplanes equipped with detecting devices over Germany to try to find out if there were any other transmissions which could be used to locate the bomber aeroplanes we were sending nightly to Germany in ever increasing numbers. It was found that the Germans were installing a radio location defence system based on their 2-metre (150 megacycles) sets in France, Belgium, Holland , Denmark and Norway. This would cover all the aerial approaches to Germany. It was also discovered that they had two other sets which were working on 500 megacycles, 60 cm wavelength and on 600 megacycles, 50 cm wavelengths, both with a range of about 15 miles. This was disconcerting because we had not so far developed jamming equipment for centimetre wavelengths which were being used for German night-fighter control. It was also discovered that German U-boats were being fitted with radio equipment which would detect the transmission of the 1.5 metre Air to Surface Vessel (ASV) Mark II sets fitted to Coastal Command aeroplanes. By detecting these transmissions the U-boats would have time to dive and so reduce the chance of being attacked. However, although they would escape being attacked they would have less time to re-charge their batteries, thus reducing their effectiveness and range.

Since the U-boat receivers would pick up the 1.5 metre transmissions coming from Coastal Command aeroplanes before the radio location set operators had located the U-boat on the surface, the number of U-boat sinkings began to fall. Constant patrolling of the routes used by the U-boats coming and going to their French bases meant that the U-boats had to submerge for much longer periods, and when they came to the surface they were always in danger of being attacked. This was the beginning of the development of ECCM (Electronic Counter-Counter Measures) – both British and German electronics specialists tried to get and keep ahead of their opponents.

By August 1941 any aeroplane, either friend or foe, coming in over the east, south and west coasts could be tracked all the time and any friendly aeroplane which had not switched on the Friend-Foe Indicator was in grave danger of being shot down, which happened in a few instances. There were some night bomber attacks by the German air force, but they were mainly in the London area and coastal towns; so far these had not been on the scale of 1940. Even so, the Germans were losing between 10 per cent and 15 per cent of their attacking force on every raid. This was of little consolation to the people on the receiving end of the German bombs and there was a general demand for the RAF to bomb German cities and towns as well as the military targets which they were already bombing.

A large area of Russia had been overrun by the German army who were now on their way to Moscow. It appeared to us that there was very little to stop them achieving their aim. At this time there was not much that we could do to help the Russians ·since we were still very short of munitions and all the arms that we were producing were being used in North Africa to stop the German and Italian armies from invading Egypt and the Middle East. We had, however, perfected a beam bombing system which would improve the accuracy of the Bomber Command night attacks. The first of these attacks was against the towns in the Ruhr area of Germany and the results were very encouraging. The number of aeroplanes lost was at an acceptable level and the attack marked the beginning of the destruction of German cities and towns by Bomber Command using electronic devices for locating and bombing targets accurately. The Germans had tried out this method of locating and bombing targets in 1940 but had been thwarted by the use of jamming equipment. As far as we knew, they had not attempted to jam any of our transmissions and this applied to the beams we were now using to bomb their cities.

Towards the end of August 1941 Dr Dodds and Mr Ludlow came into our office and told Mr Chadwick and me that we had received an initial order for 30 airborne radio location sets which were similar to the Air Intercept Mark VIII sets which we were now producing. These new sets were to be called 'Panoramic Transmitters/Receivers' and would be operated on a 10 cm wavelength, 3,000 megacycles, with an initial pulsed output of 5,000 watts. A parabolic dish, similar to the one used on the AI Mark VIII sets, was to be used, but the aerial would rotate at the same speed as the deflection coils around the neck of the cathode-ray tube, in a similar manner to those used on the Plan Position Indicators at about four revolutions per minute. The sets were to be designed so that glass triode valves or magnetrons could be used as the oscillator. Outline drawings of the valves and magnetrons to be used were given to us by Dr Dodds. We were to design and produce the power units, oscillator/modulator units, aerial coupling units, the cathode-ray tube and control housing and the parabolic dishes. The aerial turning mechanism, receiver and cathode-ray tubes would be supplied by others.

These Panoramic Transmitters/Receivers were the first airborne Plan Position Indicators to be made. Initially they would have a range of about 25,000 ft and cover an area of about ten miles' diameter, depending on the height of the aeroplane. Whilst we called these airborne PPI sets by the name of Panoramic Transmitters/Receivers, they became known as 'H2S sets' by Bomber Command and 'ASV Mark III sets' by Coastal Command, and since the magnetrons were still at the experimental stage, an initial order would be supplied with glass valves and would be used for trials by Bomber Command. It had also been decided that any aeroplane fitted with magnetrons must not leave the United Kingdom even for trials. It was very important that the Germans did not learn about the magnetron by recovering one from a crashed Bomber Command aeroplane.

As the supply of magnetrons improved the transmitters would be fitted to Air to Surface Vessel (ASV) Mark III Coastal Command aeroplanes to help them in their battle with the German submarines. During trials using the Panoramic Transmitters/Receivers it was discovered that the picture displayed on the cathode-ray tube of the area the aeroplane was flying over showed up the dividing line between the water and land very clearly. Furthermore, ships stood out very clearly against the surrounding sea. This meant that German towns and cities near rivers, canals, lakes or docking areas could be located at night and in fog, as could submarines on the surface in similar conditions.

Bomber Command were using the beams to bomb targets in Germany but they could only reach the western part of the country – the eastern parts could not be attacked using the beams because of the distances involved. This would change when the Panoramic sets which we were now designing and producing were fitted in Bomber Command aeroplanes. Also, when these sets were fitted in Coastal Command aeroplanes the detection and destruction of the U-boats would be increased with a consequent reduction in the number of Allied ships sunk.

So far in World War II all the power required for all the electronic devices being fitted in aeroplanes had come from the 24 volt battery which was charged by a direct current generator driven by one of the aeroplane's engines. Various pieces of equipment had been used to change the direct current supply to alternating current which was required for the power supplies to the electronic units. A decision was made to fit all the new aeroplanes with an alternating current generators as well as the direct current generator normally fitted. These new generators would be driven by the engines and would have an output of 1,500 watts maximum (at 100 volts) and a frequency of 1,000 cycles per second, and would power all the electronic units in the aeroplane. The reason for increasing the frequency of the supply from 500 cycles per second – which had been used so far – to 1,000 cycles per second was to reduce the size and weight of the transformers, condensers and chokes used mainly in the power supplies. We were also able to standardise the power supply units and reduce the number of different types used.

Some of the new aeroplanes being produced were being fitted with all the new electronic devices available and were to become a pathfinder force for the accurate location of targets. They would also be able to mark the targets with coloured parachute flares which would be the targets for the main bomber force. Other aeroplanes were fitted with radio detectors to detect the German defence system radio transmissions. These aeroplanes were also fitted with jamming equipment and would fly with the main bomber stream and use the jammers to confuse the German defence system and their night-fighter communications.

Towards the end of 1940 we had revealed the secrets of our Chain Home defence system to the Americans and Canadians and as a result there had been increased cooperation in the design and production of radio components and also an interchange of ideas regarding radio location equipment. We had gained by this exchange and this showed in the amount of radio parts we were now receiving; they had gained because the basic designs had been completed and presented to them in a finished form. As a result the Americans decided to install some radio location sets in and around Pearl Harbor in Hawaii and by October 1941 a number of sets were working and being used to train operators. The sets were mobile and would be similar to our Coastal Defence sets and they would have an accurate range of about 100 miles.

At about 7.00 a.m. on the morning of 7 December 1941, an operator in the Pearl Harbor area who was watching the cathode-ray tube display saw a large number of blips at a distance of about 120 miles and, after consulting the plotter, they both came to the conclusion that what they were seeing was obviously ground clutter. They kept watching and at about 7.15 a.m. they decided to tell the officer in charge that there was a lot of air activity about 100 miles away. The officer in charge said that there could be something wrong with the set or it would be some U.S. Navy aeroplanes on manoeuvres and they could forget it. They continued to watch and by 7.40 a.m. they could see blips on the cathode-ray tube about 40 miles away. At 7.50 a.m. the first Japanese bombs fell on Pearl Harbor. The officer in charge was quite correct – what the observers had seen and reported were Navy aeroplanes on manoeuvres, but it was the wrong Navy. The fatal mistake was to assume that the blips seen on the cathode-ray tube were friendly aeroplanes. If some of their own naval aeroplanes had been fitted with Friend-Foe Indicators the observers would have known that what they were seeing had to be enemy aeroplanes. Consequently the defence systems would have had about an hour to prepare.

In 1939 we in Britain had been more fortunate. The need for Friend-Foe Indicators was shown up when some of our own aeroplanes were fired on when returning from a reconnaissance flight over Germany. By July 1940 most of our fighter aeroplanes had been fitted with FFI units which responded to the Chain Home radio frequencies. We had about nine months' preparation before our defence system was really tested in August, September and October 1940. As a result of the Japanese attack on Pearl Harbor and the entrance into the war of the USA, the German U-boats were able to operate up to the North American Atlantic coast and there was a marked increase in the number of ships sunk.

During 1941 we had developed some High Frequency Direction Finding (HF/DF) receiver which could be tuned to the radio frequencies used by the U-boats for talking to other U-boats. Some of these HF/DF receivers were land-based and some were fitted to convoy escort ships. When a U-boat transmitted any signal by its radio the HF/DF receivers could pick up the direction the signal was coming from and also its strength. If the signal was picked up by two or more HF/DF receivers the position of the transmitting U-boat could be fixed and, better still, if another U-boat replied both U-boats could be located. Instead of operating as single units the U-boats had now been arranged in packs so that when a convoy of ships was found they could attack together. However, to keep in touch with each other the U-boats had to use their radio transmitters much more than they had previously. As a result, the HF/DF receivers had more chance of picking up signals.

It was now January 1942 and World War II had spread to Asia with the Japanese overrunning South-East Asia with very little to stop them. The German advance into Russia had been slowed down mainly by the weather, and the German and Italian advance in North Africa had been stopped. Night bombing by the German air force of some of our coastal towns and cities was still being carried out when the weather was suitable, but the number of their aeroplanes being shot down was increasing. Bomber Command was stepping up its attacks on German targets by using the beams system. The Germans did not seem to have any method of defeating it and consequently up to that point our losses had not been very high. The demand for all types of radio location and jamming equipment for use on land, on ships and in aeroplanes was increasing. In our factory alone there were many thousands of people producing, assembling and testing the various types of sets and the development was still all very secret.

On 12 February 1942 three German battleships which had taken refuge in the port of Brest in France made a high-speed dash through the straits of Dover to their home ports in Germany. As a result of the German battleship sorties there was a demand to know how this had come about and why our defence system had failed to detect the ships in their dash into the North Sea.

At about 8.00 a.m. on the morning of 12 February 1942 the Chain Home station at Pevensey reported some aeroplane activity just off the French coast in the Le Havre area at about 5,000 ft. This continued to about 10.00 a.m. At about 11.00 a.m. the Coastal Defence radio location Type 271 sets sited near Hastings located surface ships about 40 miles away. By this time the air activity had reduced. At about 11.30 a.m. it was very obvious that a number of German battleships with a strong escort were going through the Straits of Dover. The battleships and their escorts were attacked by the long range guns at Dover, by motor torpedo ships and by RAF bombers and some of the battleships and escorts were damaged. It was thought at the time that the Germans had developed some jamming device which they had used to jam the various radio location installations we now had along the south and east coasts. What had probably happened was that the first reports of air activity off the French coast were assumed to be the German aeroplanes forming up to make an attack somewhere along the south coast. In fact it was the air escort for the battleships. The air escort was withdrawn before the battleships reached the Straits of Dover. To the operators watching the cathode-ray tubes of the Chain Home sets and the plotters tracking the movements of all the aeroplanes it would have seemed that the German air force had decided not to attack and had gone home. Only when the Coastal Defence sets had located and tracked surface ships was it realised that what they had in fact located were German battleships. At this stage of World War II the Germans did not have any jamming equipment which would have been effective against the Chain Home transmitters working on 10 – 13 metre wavelengths and the Coastal Defence transmitters working on 10 cm wavelengths.

Chapter 14

Jamming the Enemy Defence System

ON 15 FEBRUARY 1942 Singapore was overrun by the Japanese army. Bomber Command was now making many heavy attacks on targets in Germany but the number of British aeroplanes being damaged and shot down was increasing. This was due to the increased use of radio location equipment by the German defence system which enabled their night-fighter aeroplanes to be directed onto the bomber stream. Whereas we knew the locations of a large number of their defence sites and the radio frequencies their sets operated on, we did not know in detail how their sets were constructed, what valves, cathode-ray tubes and pulse generators they had developed and how good their receivers were. On 27 February 1942 a commando raid was carried out on a German Würzburg site at Bruneval just north of Le Havre on the French coast. The raid was very successful and a number of radio components and other parts were brought back for examination. As a result their operating wavelength of 50 cm and range of 12 miles was confirmed.

The information made available as a result of the raid enabled us to decide what type of jamming equipment to produce which would be capable of jamming all the Würzburg radio location sets being used in the German defence system. We already knew that for long range (60 miles) surveillance the Germans were using radio location sets which they called 'Freya' operating on a wavelength of 2.5 metres (120 megacycles). These transmissions would also have to be jammed. It was most important that once we started to use jamming devices against the German defence system they must be very effective.

Various trials were carried out on all the radio frequencies used in the German defence system and the best results were obtained by using radio transmitters on 2.5 metres suitably modulated to distort the 'Freya' receiver. These jamming transmitters would be mostly airborne but some would be shipborne. To jam the 50 cm wavelength Würzburg sets it was found that strips of very thin metal foil measuring about $^1/_8$ in. wide and about 10 in long ($^1/_2$ wavelength) dropped from Bomber Command aeroplanes (which had been fitted with centimetre wavelength detection equipment) would confuse the Würzburg operators. These new devices were put into production so that when we came to use them the German defence system would be so overwhelmed that it would not recover. Our losses in aeroplanes would consequently be reduced.

By March 1942 a large number of Bomber Command aeroplanes had been fitted with the latest electronic navigation equipment. The crew of each aeroplane fitted with this new navigational aid were able to determine the plane's position very accurately on an electronic grid giving its latitude and longitude at any given time. This new system had a maximum range of about 280 miles, and since the aeroplane's position relative to the target was being continuously updated, the aeroplane could approach the target from any direction. As a result of this ability to approach a given target from a number of directions the German defence system was unable to predict the possible target and only knew which one had been selected when coloured marker flares were dropped and bombs began to fall. There was also another big advantage: by arranging for the aeroplanes to approach the target from different directions and at different heights, the time taken for the attack could be reduced and the German defence system had less time to react. In this it was hoped that the new system would reduce Bomber Command's losses in their nightly bombing campaign. Since it used a number of radio beams, it was very difficult to jam. In May 1942 the new bombing system was used to attack Cologne. Over 1,000 aeroplanes were used in the attack which was very concentrated and of short duration. The German defence system was completely overwhelmed.

The production of magnetrons was now increased and in July 1942 a decision was made to stop using the glass triode valves for the Panoramic transmitter/receivers we were now making and replace them with the 10 cm wavelength magnetrons. The magnetrons were made entirely of metal and did not require as much cooling as the glass valves. They did, however, require about 10,000 volts DC on the anode cavities. This gave a pulsed output of about 10 kilowatts which was double that of the glass valves. Trials using the coastal Command ASV Mark III fitted with magnetrons had been very successful, so it was decided to give priority to Coastal Command aeroplanes in their battle with the German submarines they would not be able to detect the 10 cm transmissions from the new ASV Mark III sets which were now being fitted and the submarines would be located and attacked while on the surface at night and in fog.

The German army had overrun a large part of Russia and the Japanese army and navy had conquered a large part of South-East Asia and also a number of islands in the Pacific Ocean. In North Africa the advance of the German and Italian armies had been stopped at El Alamein in Egypt, but the Battle of the Atlantic did not seem to be going in the Allies' favour. In October 1942 the British forces at El Alamein launched an attack on the German and Italian armies with the object of removing the enemy from North Africa. On 8 November an Allied landing took place in North Africa behind the German and Italian armies. To help make this landing possible some of the ships where fitted with Ground Control Intercept (GCI) sets and a number of night-fighter aeroplanes fitted with Air Intercept (AI) Mark IV sets were used to defend the area around Algiers from nightly air raids by the Italian air force. This was the first time the GCI and AI Mark IV sets had been used together overseas.

As a result of the commando raid on the German Würzburg site and other observations of the German fighter control system, a number of fighter direction stations were installed and equipped with a new type of radio location telescope with a parabolic aerial about 60 ft in diameter working on a 50 cm wavelength (600 megacycles). It was designed to have a very accurate direction finding system. These new sets were designated to watch the air space above enemy-held territory for any activity by the German air force and to locate where it was taking place. From the information gained, RAF Fighter Command were able to make daylight attacks on German aeroplanes both in the air and on the ground. Such attacks were called 'fighter sweeps' and were reported in all the newspapers of the day.

The main object of these fighter sweeps was to destroy the enemy air force and prevent the Germans from moving a large number of their fighter aeroplanes to the Russian front; they would have to keep a large number of aeroplanes to defend their western front.

By January 1943, all the radio location equipment we had been designing and producing was having a big effect on the progress of World War II. So far, most of the equipment we had made was mainly for defence purposes but the latest equipment was being used to attack the enemy. As a consequence of the use of the HF/DF receivers for locating U-boats and the use of the ASV Mark II and ASV Mark III radio location sets by Coastal Command for attacking them, the number being sunk was increasing. U-boat bases in France were being attacked night and day. In April 1943 it was claimed that 35 U-boats had been sunk within a single month. There was a big reduction in the number of Allied ships sunk and the Allies were beginning to win the Battle of the Atlantic.

As a result of the increasing number of attacks on the U-boats, the Germans designed a device called a 'Schnorkel' which they started fitting to their submarines. This enable the submarine to remain just below the sea level with the Schnorkel protruding above the sea level. In this manner the submarine's diesel engines could be run to charge the batteries, the air for the engines was drawn down the Schnorkel tube and the exhaust gases were blown out of the tube. When these Schnorkel tubes were fitted to the submarines, Coastal Command and the Navy had more difficulty in locating and attacking them even in daylight. The ASV Mark II sets working on a 1.5 metre wavelength would not locate such a small object as the Schnorkel tube against the background of the sea and the location range for type ASV Mark III working on a 10 cm wavelength would be reduced to about three miles. However, the HF/DF receivers used for listening to the radio transmissions between U-boats and their bases in France would not be affected by the Schnorkel tubes and these would play an increasingly important role in their final defeat.

Now that production of magnetrons had been increased we were able to produce more Panoramic (H2S) and ASV Mark III sets for use in Bomber Command and Coastal Command aeroplanes. A decision was made to try out the Panoramic (H2S) sets which were fitted in Bomber Command aeroplanes. They were to be used for locating targets in Germany which could not be reached using the beam system. It was accepted that sooner or later the Germans would recover a magnetron from a crashed aeroplane, but it was the general opinion that it would take them at least two years to produce them in quantity. With the attacks on German industry on the increase it would probably take them much longer. The type ASV Mark II working on 1.5 metre wavelength was withdrawn and modified and became Type 286 which was used by the Royal Navy. All

Coastal Command aeroplanes were fitted with ASV Mark III sets.

Hamburg, the naval base in north-west Germany, was being used for the production of warships and U-boats and was well defended with the best equipment the Germans could produce and consequently it had not been attacked a great deal so far. Reconnaissance flights by Bomber Command pathfinder aeroplanes fitted with Panoramic (H2S) sets had shown that the waterways and docks could be located at night and in fog and by using the latest radio-jamming equipment the German defence system would be nullified. However, using the new jamming equipment would reveal its effectiveness to the Germans who could then try to jam our own defence and attack systems. Fortunately our radio location sets were on wavelengths from 15 metres to 10 centimetres – with such a range of wavelengths it would be difficult to jam them all. The Germans tended to use 2 metre and 50 and 60 centimetre wavelengths and so their systems were more susceptible to jamming.

On the night of 24-25 July 1943 about 900 bomber aeroplanes of Bomber Command set out for Hamburg. They were led by pathfinders fitted with the latest Panoramic (H2S) sets for locating the targets and marking them with coloured flares. Together with the pathfinders were aeroplanes fitted with detectors and jamming radio sets to jam the 2 metre wavelength defence transmissions. There were also other aeroplanes fitted with devices for detecting the 50 and 60 centimetre transmissions. These aeroplanes carried bundles of metal foil strips which were thrown out of the aeroplanes when the centimetre transmissions were detected. All the aeroplanes which followed them carried high explosive bombs and incendiary bombs. The raid was so concentrated that it took about one hour for all the aeroplanes to drop their bomb load on Hamburg. The defence system was completely overwhelmed which was confirmed by listening to the two-way radio communications between the German night fighter pilots and their ground controllers. Over the following days the U.S. Air Force attacked Hamburg by daylight, while Bomber Command returned each night until 29 July. These air raids on Hamburg were the most devastating of any air attacks so far in World War II. Because of the weight of the bombs dropped in such a short time and the follow-up raids, the fires could not be controlled and these burned for days after the attacks.

This was the most successful series of air raids so far undertaken by Bomber Command. Many of the new four-engined bombers which could carry a larger bomb load took part in the raids which had been concentrated into a short space of time. The Allies' losses had not been very great and they had shown the Germans that even their best defended towns and cities could be attacked and destroyed. Some of the bombers lost in the raid were fitted with the Panoramic (H2S) sets and the jamming equipment. When the Germans found some of these crashed aeroplanes they would have very quickly learned the secrets of the new devices being used to locate targets and confuse the German defence system.

Most of our work was now concerned with the modification of existing types of radio location equipment and with keeping drawings and specifications up to date. The new aeroplanes now being produced to carry all the up-to-date radio location, radio detecting and jamming equipment were being fitted with alternators with a maximum output of about 2,000 watts. A new magnetron working on a 3 centimetre wavelength (10,000 megacycles) had been developed and was being used to replace the 10 centimetre magnetrons as it became available. This was the Panoramic ASV Mark VII set used in Coastal Command aeroplanes. As more 3 centimetre magnetrons were supplied, they were fitted to Panoramic H2S sets and became Panoramic H2S Mark III sets, used by Bomber Command pathfinders.

These new sets showed a much clearer image on the

cathode-ray tubes than any previous Panoramic set and they could distinguish between built-up areas such as towns and cities and open country, even when the aeroplane was flying at maximum height at night and in fog. These sets would locate and pinpoint the German industrial areas so that the following Bomber Command aeroplanes could destroy the manufacturing centre of Germany. So far the Germans had not developed any electronic equipment to detect or jam these new high speed aeroplanes fitted with the Panoramic H2S Mark III sets. Likewise the Panoramic ASV Mark VII fitted to Coastal Command aeroplanes could not be detected by the U-boats and, since the image on the cathode-ray tube was much clearer, the German submarines were being attacked and sunk in ever increasing numbers.

As a result of all this effort by Coastal Command and the Navy, shipping losses in the Atlantic Ocean began to fall dramatically. By December 1943 the U-boats were beaten. Although some shipping was sunk after this date, losses experienced were never on the scale from 1940 to early 1943. 1943 was a turning point for the Allies, although it did not appear so at the time. The Russians had stopped the German advance at Stalingrad and were now driving them back towards Germany. Italian and German armies had been driven out of North Africa. Sicily had been taken in August 1943 and an invasion of the Italian mainland had taken place. Bomber Command had made some very heavy attacks on Berlin and other towns and cities in East Germany. The U.S Air Force, now backed with fighter cover, were attacking targets in daylight in Germany and the occupied countries. As a result of these attacks, particularly those on Berlin, the German Air Force renewed their night bombing of London and other southern towns and cities, but these attacks were never on the same scale as those of 1940 and 1941.

By the end of February 1944 German losses in men and aeroplanes were so great that the night bombing could not be sustained and they reverted to hit and run raids with single aeroplanes. As a result of finding out more about the German defence system and how it operated, some new airborne jamming equipment we had designed was now in production and being fitted in new high speed aeroplanes of Bomber Command. This new radio system we called 'Jostle', and it was designed to blot out all the radio frequencies used by the Germans in the defence and radio communications used by their armed forces. It was the most powerful jamming device that had been made so far in World War II and could cover an area of about 200 square miles when an aeroplane was flying at its maximum height. It would cause confusion among the Germans who would come to distrust what they were seeing on their cathode-ray tubes and hearing on the radio receivers used by the defence controllers.

Chapter 15

German V1s and V2s

THERE WAS NOW quite a lot of talk about a second front in the newspapers and some of them tried to show the generals how it could be done with diagrams and articles written by armchair military strategists. The Allied advance in Italy had come to a halt at Monte Cassino but on 18 May 1944 the town was captured and the advance continued towards Rome. Bomber Command, Fighter Command and the U.S. Air Force were now involved in a great effort both by night and day to gain complete air superiority over the German air force along the French and Belgian coastline by attacking airfields, defence sites and known German radio location installations in these areas. From what the newspapers were saying, something was going to happen soon, but what it was and where it would take place, nobody seemed to know.

On 4 June 1944 the Allies captured Rome and this was reported in the newspapers. Then on 6 June 1944, the BBC broadcast unconfirmed reports that an Allied landing in France had taken place at about 6 a.m. that morning. This was D-Day, the largest airborne and seaborne invasion ever undertaken against well defended beaches and coastlines. Very soon we were getting reports in the newspapers and on the radio of what was going on in Normandy. According to the reports the invasion seemed to be going as planned.

On 13 June 1944 the first pilotless aeroplanes, the

German V1s, landed in and around London. Whilst some intelligence reports had indicated that the Germans had developed a flying bomb, we did not know very much about how it worked, its range or the size of its bomb load. Since the V1s flew at a constant speed and at a predetermined height, it was fairly easy for the Chain Home defence system along the south and east coasts to locate and track their progress towards the targets. The speed of the V1 was about the same as that of the latest RAF fighter aeroplane; it had a long flaming exhaust shooting out of the back end and it would therefore be seen both at night time and in daylight. There were about 100 V1s being aimed at targets in and around London every day. The RAF fighter aeroplanes tried to shoot them down over the sea and those they missed were fired at by anti-aircraft guns inland. But even when the V1s were successfully shot down they caused a lot of damage.

Quite a lot of the anti-aircraft batteries had been fitted with the latest Gun Layer Mark III radio location sets with automatic tracking, and early in July 1944 a decision was made to move these anti-aircraft batteries to sites along the coast. These sites were decided on by using information gathered by the Chain Home radio location sites along the south coast. Within a few days, these anti-aircraft batteries were shooting down the V1s as they came over the coast and, since the Gun Layer Mark III was designed to give the best results when the target was flying at a constant speed and at a known height, most of the V1s were being destroyed either by RAF fighter aeroplanes or the anti-aircraft guns.

Another device had been designed and produced which was a spin-off from developments in radio location: this was known as the 'Proximity Fuse' or the 'Variable Time Fuse' which was now being used in the projectiles fired from the anti-aircraft guns. When these new fuses were used the projectile did not have to hit the target – a near miss would be enough to explode both the projectile and the target. Even if the target was not destroyed it would be damaged.

The tracking information which had been collected by the Chain Home sites was put to further use in so far as the points along the French coast where the V1s were being launched could be noted. These details, combined with other information that had been gathered regarding the performance of the V1s, meant that the launching sites could be located. Bomber Command and Fighter Command kept up day and night attacks on these sites and by early September 1944 the V1s had been more or less defeated. Quite a lot of the launching sites in France had now been overrun by the Allied armies.

The German air force started to launch the V1s from bomber aeroplanes flying over the North Sea, releasing the V1s when the target came within range. Such attacks were made on towns and cities in the Midlands and the North but they were not very accurate and not on the same scale as the attacks on London and the southern towns. The Germans launched an estimated 8,000 V1s and about 5,000 were destroyed by the RAF fighter aeroplanes or the anti-aircraft batteries.

There was a great sigh of relief throughout the whole of Great Britain when it was announced in early September 1944 that the V1s had been defeated. This relief was short-lived.. On 8 September 1944 the first V2 rockets landed in London. We had no defence against these rockets except to attack their launching sites which were mainly in Belgium and Holland and also in Germany. The Chain Home radio location sets along the east coast were able to track the V2s as they rose from their firing pads into the heavens, but the tracking could only be done over the first few seconds of their flight because the V2 rockets went very high before returning to earth and the target. A combination of constant surveillance of the Chain Home cathode-ray tubes and the reconnaissance flights by the RAF led to the location of the launch sites which were then attacked. The V2s were about

50 feet long and weighed about 15 tons. Their speed at impact on the target was faster than that of sound so the explosion of the warhead (followed by a noise like an express train) was the first indication that a V2 had landed. There were about 1,200 V2s aimed at London and about 500 reached their target. The others came down over a wide area causing great damage wherever they fell. The last one fell on 27 March 1945. The V2 rocket became the basis for the American and Russian space programmes. In 1944 the Germans were much further advanced in rocket propulsion than the Allies.

During the Allied landings in Normandy and for some time after, all the efforts of the RAF and the U.S. Air Force went into assisting the ground forces to overcome the German armies in France. By January 1945 Paris had been taken; the Allied armies had reached the Rhine in Germany and the Russians were about 50 miles from Berlin. It was decided that Bomber Command should resume their night bombing attacks while the U.S. Air Force should restart their daylight bombing of targets in Germany.

On the night of 14 February 1945, Bomber Command pathfinders, using the latest Panoramic H2S Mark III sets, marked out Dresden with coloured flares for the following bombers to attack. Some of the aeroplanes which flew with the bomber stream and others used for diversionary attacks were fitted with the latest jamming devices to confuse the German defence system. This was the first time Bomber Command had attacked Dresden. It was the most devastating attack on any city so far in World War II. On 28 April the German armies in Italy surrendered. The Russians had encircled Berlin and on 7 May the German forces in Germany surrendered at Rheims. The war in Europe was over. 8 May became, VE day but Japan still had to be defeated.

At the time we did not know how much the Japanese knew about radio location. It was possible the Germans had given the Japanese details of the equipment they were producing and also as much information as they themselves had about the devices the Allies were using. We had only revealed the secrets of our equipment to the Americans and some Commonwealth countries – we had not revealed any details to the Russians. In the Far East the Americans were overcoming Japanese resistance by using their naval forces to cut off supply ships to the various islands in the Pacific Ocean which the Japanese had overrun. As a result of the American success and their approach to the Japanese mainland, the Japanese air force began to form kamikaze attacks on American warships. The kamikaze pilots just dived their aeroplanes straight at the target they had selected and both attacker and target would be destroyed. The Americans lost quite a number of ships in this way. A new type of radio location set had been designed and produced for Royal Navy ships which was similar to the Gun Layer Mark III with automatic tracking (used by the anti-aircraft batteries on land). These new sets were Naval Types 267 and 268. The American version, Type SCR584, was designed for defence against air attack and dive bombing in particular. The advent of these new automatic tracking sets combined with the proximity fuses now being used (described earlier) ensured that the kamikaze pilots were often shot down before they could reach their targets.

On 6 August 1945 a single atomic bomb was dropped on Hiroshima in Japan. On 9 August 1945 a similar bomb was dropped on Nagasaki. Both these cities were completely destroyed. These were the most devastating attacks of World War II and as a result, the Japanese surrendered on 14 August 1945. On 2 September 1945 World War II was over. The end came much quicker than anyone expected; there were still thousands of people producing all kinds of electronic equipment for the war effort.

Now that the war was over, it was decided that radio location, or radar sets as they would now increasingly become known, were only a wartime requirement and there would be very little demand for them in peacetime.

As a result of this decision the demand for new equipment diminished. We completed outstanding orders very quickly and the workforce which had been producing the radio equipment was diverted onto making electrical components for peacetime use. Mr Chadwick reached retirement age and I was given the job of sorting out all the drawings, specifications and correspondence which had accumulated since 1936. If in doubt as to which pieces of paper to incinerate, I was to check with Dr Dodds or Mr Ludlow – they were all still very secret.

From 1936 to November 1945 I had been involved in the design, specification and manufacture of most of the transmitters used on radar sets used in World War II. I knew nothing about any other type of electrical equipment. I was still only 29 years of age, so I decided to leave the company and try something different. Later Dr Dodds was awarded the OBE for his contribution to the development of radar, but I don't think Mr Ludlow received any award for his effort. The four 350-foot-high transmitter masts were still standing at Danby Beacon, North Yorkshire in 1957, those near Dover also remained in 1959. I do not know how long the Danby Beacon masts were in position nor do I know when the Dover masts were dismantled. This information could be obtained if it was necessary to know. RAE at Farnborough would know and no doubt have some photographs. I believe the Germans photographed them from France at the end of 1940.

Appendix

TOWARDS THE END of the 1920s and in the early 1930s it began to be realised in scientific circles that radio waves could be reflected and detected in a similar manner to both visible and invisible light waves. There was soon a great effort to predict weather by locating cloud formations and, in particular, storm clouds. Much effort went into trying to find out the directions these clouds were coming from, their height and their speed of approach. It was also realised that there would be a great advantage if a means could be devised to enable ships to locate other vessels and objects such as icebergs in foggy conditions, particularly in the North Atlantic.

But the greater urgency in the western world was to be able to detect aeroplanes. Bomber aeroplanes being developed by the western countries were becoming faster and bigger, and could carry quite a heavy bomb load, and the general opinion in the early 1930s was that there was no defence against a determined bomber attack. The only possible defence would be a continuous watch using fighter aeroplanes which would be very expensive, and even then, some of the bomber aeroplanes would be likely to get through the fighter screen. There seemed to be no solution to the problem of the bomber aeroplane.

This was the position when Mr Watson-Watt, who had been engaged on predicting weather by the use of radio waves, was asked if he could develop equipment that would locate aeroplanes flying at various heights and speeds and give their distance, height and position in relation to the transmitter aerials. It was a formidable task and he was not helped by some of the so-called experts of the time whose ideas and suggestions were of little use but had to be explored. Mr Watson-Watt carried out trials using the best glass valves that could be obtained and from these trials he was able to calculate the amount of power and the best frequencies required to produce a radio location system. He established that the best wavelengths were 1 metre (300 megacycles) or less than 1 metre with about 100 kilowatts radiating from the transmitter aerials.

In 1934 there was no radio equipment available which would meet with these two basic requirements. There were, however, some quite powerful transmitters working in 50 metres (6 megacycles) at Daventry used by the GPO for overseas radio communications, and there had been reports from the engineers at Daventry that when aeroplanes were flying in the area, the radio transmissions were affected by their passage. Mr Watson-Watt arranged for tests to be carried out using the Daventry transmitters and an aeroplane flying at known distances and heights from the transmitting aerials. The results were the best so far achieved. The valves being used at the Daventry transmitting station were those developed by Dr Dodds and Mr Ludlow in the Valve Lab at Met-Vick. It was now about February 1935.

As a result of these trials and some modifications to the valves, it was found that the valves would be made to oscillate at a wavelength of 10 metres (30 megacycles) with an output of at least 60 kilowatts for each valve. Armed with these facts and figures, Mr Watson-Watt decided that a suitable defence system to detect and locate aeroplanes in daylight, at night and in fog could be developed using the radio valves developed by Dr Dodds and Mr Ludlow. As a result Met-Vick was given an order for one transmitter station which consisted of two transmitters, one on the air, the other on stand-by. Each transmitter consisted of two cubicles, one for the oscillator and the other for the modulator, and a control desk for controlling both cubicles and monitoring the output to the aerials. The transmitters were to operate on wavelengths of 10 to 15 metres (30 to 20 megacycles) with a wave change system so that each transmitter could be operated on one of four wavelengths which could be changed in about five minutes. The reason for the wave change system was because there was the possibility that a single frequency system could be jammed. To jam four frequencies would be very difficult.

The output of each transmitter would be about 300 kilowatts. Modulation of the transmitted wave would be effected by using the 50 cycle per second frequency of the national grid supply. It was thought that with this method of modulation anyone on the other side of the North Sea who picked up the transmissions would simply think there was something wrong with the national grid network.

The completed transmitters were to be delivered to Bawdsey in Suffolk as soon as possible. This is where I became involved in the design and production of radar equipment in the Valve Lab Research Department at Met-Vick. It was February 1936 and I was 20 years old. The receivers and aerials would be supplied by other manufacturers. The reason for this was that the whole operation had to be conducted in great secrecy and it was also important that no single manufacturer should have a monopoly on the equipment being produced. Dr Dodds and Mr Ludlow did all the design work and Mr Chadwick and I did all the drawings, specifications and manufacturing instructions for the transmitters and control desks. They were designed for 'push-pull class B operation' in their initial arrangement and could detect and locate aeroplanes at about 80 to 100 miles from the aerial.

In 1939 they were modified to 'pulsed class C operation' which increased the effective power to the aerials to about 650 kilowatts and increased the detection range to 150 –

180 miles. This pulsed arrangement was a big breakthrough in radar technique and was used on all the sets that followed. We shipped the transmitters for the first link in the Chain Home (CH) defence system to Bawdsey in July 1937. Altogether we supplied transmitters for 22 Chain Home (CH) sites from Ventnor to Dover on the south coast and Dover to the Orkney Islands on the east coast. There were 88 cubicles and 44 control desks in all and most of these were working by September 1939. At this time the project was all still very secret.

During trials at Bawdsey in 1937 and early 1938 when a number of aeroplanes were used to stage mock attacks, it was found that low flying aeroplanes could not be detected by the Chain Home receivers. As a result we were asked to design and produce transmitters for Chain Home Low (CHL) stations which were initially designed to operate on a 7 metre (43 megacycles) wavelength. This was very quickly changed to 1.5 metre (200 megacycles) as new and better glass valves were developed. We were also asked to design some mobile sets for Coastal Defence (CD) also on a 1.5 metre wavelength. This band of wavelengths was used for all kinds of radar sets in World War II.

At the outbreak of war in September 1939, every single aeroplane was being detected and located by the Chain Home stations and there was nothing to distinguish friend from foe. Mr Watson-Watt and his team at Bawdsey designed a Friend-Foe Indicator (FFI) which we were asked to produce. We must have made several thousands of these.

By June 1940 the Chain Home defence system was working and was about to be put to the test in earnest by the German air force. The transmitters were the most powerful short wave sets made up to 1940. Their detection range was at least 150 miles and they could locate the German aeroplanes forming up over the French airfields for an attack. This gave our defences about 30 minutes to get prepared. We knew the direction from where an attack was coming and the approximate number of German aeroplanes in any formation and also the approximate height at which the aeroplanes were flying. We also knew they were enemy planes because most of our fighter aeroplanes had been fitted with Friend-Foe Indicators which meant that they showed as flashing spots of light on the Chain Home receiver cathode-ray tubes. Foes showed as steady spots of light.

Some of the Chain Home Low and Coastal Defence receivers had been installed but not all of them. At the time of the fall of France the coastline to the west of Ventnor in the Isle of Wight was unprotected but, fortunately, the Germans did not realise this at the time. This was the position in July 1940 when the Chain Home defence system was still very secret.

Air battles over the south and east coasts continued until October 1940. During this period the German air force lost between 1,800 and 2,000 aeroplanes, for which some of the credit must go to the Chain Home defence system developed by Mr Watson-Watt. The radar sets we designed and produced for the Royal Navy and Coastal Command helped in the destruction of large parts of the Italian and German naval forces. Of the 1,100 U-boats which were built, about 900 were destroyed.

After October 1940 the German air force resorted to night bombing, but with the aid of airborne (AI) and ground control intercept (GCI) sets, it became too expensive for the Germans to send their bombers even at night.

Development of the magnetron was another breakthrough. By using these we were able to design quite powerful airborne sets like the Panoramic (H2S) and the Panoramic ASV Mark III on 10 centimetre (3,000 megacycles) wavelengths. The Panoramic (H2S) helped in the destruction of German industrial war production.

Finally, jamming equipment was designed and produced which reduced the German defence system to utter chaos – when it was used on D-Day it confused the German High Command completely. In the Far East the airborne and

naval radar sets helped in the destruction of the Japanese
Navy and the Japanese kamikaze air force pilots. All this
came about because of Mr Watson-Watt's belief in his basic
ideas for the production of a defence system using radio
waves. It was also because of the effectiveness of the radio
valves designed by Dr Dodds and Mr Ludlow in the Valve
Lab at Met-Vick; they were there at the right time.

For his efforts in the development of radar and, in parti-
cular, the Chain Home defence system, Mr Robert Watson-
Watt received a knighthood and Dr Dodds was awarded the
OBE. Without the Chain Home defence system, in 1940
Fighter Command and the anti-aircraft guns around southern
England would not have been able to locate and destroy a
large part of the German air force in the way that they did.